THE PERPETUATION OF LIFE

Revised Nuffield
BIOLOGY
TEACHERS'
GUIDE 4

Published for the Nuffield Foundation
by Longman Group Limited

Longman Group Limited
London
*Associated companies, branches, and representatives
throughout the world*

First published 1966
Revised edition 1975
Copyright © The Nuffield Foundation, 1966, 1975
ISBN 0 582 04608 4

Design and art direction by Ivan and Robin Dodd

Filmset in 11 on 12 point Century Schoolbook
by Keyspools Ltd., Golborne, Lancashire
and made and printed in Great Britain
by Butler and Tanner Ltd., Frome and London

Contents

Foreword

It is ten years since the Nuffield Foundation undertook to sponsor curriculum development in science. The subsequent projects can now be seen in retrospect as forerunners in a decade unparallelled for interest in teaching and learning not only in but far beyond the sciences. Their success is not to be measured simply by sales but by their undoubted influence and stimulus to discussion among teachers—both convinced and not-so-convinced. The examinations accompanying the schemes of study which have been developed with the ready cooperation of School Certificate Examination Boards have provoked change and have enabled teachers to realize more fully their objectives in the classroom and laboratory. But curriculum development must itself be continuously renewed if it is to encourage innovation and not be guilty of the very sins it sets out to avoid. The opportunities for local curriculum study have seldom been greater and the creation of Schools Council and Teachers' Centres have done much to contribute to discussion and participation of teachers in this work. It is these discussions which have enabled the Nuffield Foundation to take note of changing views, correct or change emphasis in the curriculum in science, and pay attention to current attitudes to school organization. As always, we have leaned on many, particularly those in the Association for Science Education who, through their writings, conversations, and contributions in other varied ways, have brought to our attention the needs of the practising teacher and the pupil in schools.

This new edition of the Nuffield Biology *Texts* and *Teachers' guides* draws heavily on the work of the editors and authors of the first edition, to whom an immense debt is owed. The first edition, published in 1966, was edited by Professor W. H. Dowdeswell, organizer of the Biology project which carried out the trials in schools of the original draft materials. The authors of the first edition

were:
Alison Leadley Brown Texts I and II and Teachers' Guides I and II
C. D. Bingham Texts I and II and Teachers' Guides I and II
A.K. Thomas Text III and Teachers' Guide III
A. Ellis Texts III and IV and Teachers' Guides III and IV
A. Darlington Texts III and IV and Teachers' Guides III and IV
P. J. Kelly Text V and Teachers' Guide V

The new edition contains a preponderant part of these authors' material, either in its original form or in edited versions. They are credited among the authors of the new edition but their wider contribution in providing a firm basis for further developments must be gratefully acknowledged here.

I particularly wish to record our gratitude to Grace Monger the General Editor of this new series. It has been her responsibility to organize and coordinate this revision and it is largely through her efforts that we have been able to ensure the fullest cooperation between teachers and the authors.

As always I should like to acknowledge the work of William Anderson, our publications manager, and his colleagues, and, of course, to thank our publisher, the Longman Group, for continued assistance in the preparation and publication of these books. I must also record our debt to those members of Penguin Education who were actively involved in the preparation of the books until a late stage in their production. The contribution of editors and publishers to the work of the course team is not only most valued but central to effective curriculum development.

K. W. Keohane
Co-ordinator of the Nuffield Foundation Science Teaching Project

Preface
to the first edition

Teaching is so much a personal art that it would be wrong to suggest any single, rigid approach. At the same time we must remember that a course very largely succeeds or fails because of the way it is taught. *How* we teach is certainly as important as *what* we teach.

This course incorporates a selection of those teaching methods that appear to us to be most suitable for attaining its aims. Although it is written as a coherent whole, many teachers (initially, at least) may feel unable to adopt it in its entirety. They may lack the proper facilities, time, or experience. Moreover, they may wish to adapt it to suit their own particular circumstances.

Such considerations need not stand in the way of change. We believe, however, that where a reduced course is planned it should have the same broad structure as the text. The aims of the Project will not be achieved without proper attention to its main themes and principles. For example, the later parts of the course include studies in population problems, genetics, and physiology. But the special circumstances of a school may well lead to a particular emphasis being placed on one or more of these aspects of biology.

Where time is limited, the findings of other people can, with discretion, be substituted for class practical work. If parallel classes are following the same course it will sometimes be possible for them to exchange findings from different investigations. Moreover, demonstration by the teacher can, on occasions, be as effective in making a point as a class practical – or even more so. In some ways much can be done to overcome difficulties imposed by lack of time and facilities.

Aims of the course

These aims can be summarized briefly as follows:

a To develop and encourage an attitude of curiosity and enquiry.

b To develop a contemporary outlook on the subject.

c To develop an understanding of man as a living organism and his place in nature:
1 The usefulness and social implications of biology in relation to man's everyday needs, such as food and public health
2 The profound influence of man's activities on other organisms
3 The way in which a study of biology enables man to interpret observations that he makes in everyday life, for instance, about the distribution of plants and animals.

d To foster a realization of the variety of life and of underlying similarities among living things.

e To encourage a respect and feeling for all living things.

f To teach the art of planning scientific investigations, the formulation of questions, and the design of experiments (particularly the use of controls).

g To develop a critical approach to evidence.

h To develop the following ideas about biology as part of human endeavour:
1 Biology has been developing over many centuries: there are many unanswered questions about life; our ideas of life may change as new knowledge is obtained
2 Biological knowledge is the product of scientists working in many different parts of the world; its pursuit is international
3 It is based not only on observation and experimentation but also on questioning, the formulation of hypotheses, testing of hypotheses, and, above all, communication between people
4 Developments in chemistry, physics, and mathematics are helping us to make advances in biology.

Preparation and homework

Preparation and homework constitute an integral part of any course of teaching. As far as the Nuffield O-level Biology course is concerned, prepared work can fulfil an important role in helping to ensure that the atmosphere of enquiry which we have tried to build up during school hours is not broken every time a pupil leaves the laboratory. Inevitably, a certain amount of time out of school will have to be devoted to the writing of practical work (see 'Long-term investigations' on page xv), but we suggest that this is kept to a minimum. Following the advice of the majority of teachers engaged in the preliminary trials of the course, we have not included lists of questions at the end of each chapter. However, in developing each topic in the *Text* we have posed numerous questions to which we have frequently not supplied the answers. These can provide useful material for homework. Further suggestions are made in various sections of the *Guide*. At the end of most chapters in the *Text* we have included a section entitled 'Background reading' – usually an extension of some topic which it has not been possible to develop elsewhere. We hope this may also play a part in stimulating additional reading out of school hours. We suggest that such reading could also provide a link between the teaching of biology and English in a school; essays and discussions might well be based on some of the topics included.

Using the *Text*

In writing the pupils' *Texts* we have borne the following considerations particularly in mind.

a We have tried to avoid giving the impression that the *Text* is intended to usurp the position of the teacher in relation to the class. On the contrary, we have attempted to suggest in the *Text* and *Guide* a sufficient variety of teaching approaches to enable a teacher to develop those most suited to his own particular aptitudes and personality. Wherever possible, however, we have tried to cut down the time spent in class on purely routine activities. For instance, we have attempted to provide a logical introduction to all practical work and adequate instructions for carrying it out, in the hope that this will minimize the amount of time spent on preliminary demonstration of equipment and the detailed explanation of techniques. More of the teacher's time can thus be spent in discussion and dealing with the problems of particular pupils. Similarly, the introductory material on a topic can be read by the pupils for preparation in order to save time in class.

b The *Text* is intended to provide a clear indication of the overall structure of the course. This includes the main divisions of the subject matter; how different topics are related to one another; how the subject matter can be approached in an investigatory manner similar to a scientific enquiry; and how firsthand information (derived from the laboratory) and secondhand information (obtained, for example, from the books and film loops) can be related to one another.

c Throughout the *Text*, numerous opportunities are provided for initiating simple investigations and formulating hypotheses; questions are posed to which the answers are not, at once, given. In fact, the appropriate results or answers may well be incorporated in subsequent material, but we hope that the method of presentation does not make this too obvious when the pupils are first faced with an open-ended situation.

d Where investigations cannot be conducted on living organisms, models have sometimes been devised to simulate the structures or functions being investigated. Care is needed to keep the model situations closely related to those obtaining in the living things themselves.

e No separate laboratory manual has been produced for the course. The *Text* includes details of all laboratory work, our aim being to enable students to carry out individual or group experiments and investigations with the minimum of technical help. Where techniques requiring a number of manipulative processes are involved, the sequence of events has been summarized for easy reference in the form of flow diagrams.

Fitting the course into the school year

While the material contained in each chapter of the *Text* is developed in a logical order, there is no set sequence to the course as a whole. Apart from seasonal considerations, which are important for certain sections such as the work on some of the living organisms, it is intended that teachers should adapt the course freely to suit their own preferences and the facilities available.

Teaching and enquiry

One of the principal aims of the Nuffield Biology course has been to foster a critical approach to the subject with an emphasis on experimentation and enquiry rather than on mere factual assimilation. How can we foster this 'spirit of enquiry' in our pupils? There is no simple answer to this question, for teachers convey an attitude to their subject in a variety of different ways. Moreover, we must bear in mind that with younger pupils it is essential for them to experience a sense of achievement in their practical work. This requirement may well conflict to some extent with the presentation of a genuine experimental situation which inevitably involves some degree of uncertainty. The following suggestions may provide a helpful starting point.

a We suggest that, whenever possible, pupils should gain their initial understanding of a basic concept by being faced with a series of problems.

b The next phase involves the formulation of a hypothesis – the point when curiosity or observation changes to constructive questioning.

c Where possible, we suggest that deductions from experimental work should always include suggestions for a further stage in the enquiry, thus fostering the idea that one experiment paves the way for the next.

d We sometimes forget that work done outside schooltime is just as much an integral part of a course as the teaching itself. Problems and test questions set in a spirit of enquiry can do much to maintain and enhance attitudes acquired in the laboratory.

e One of the advantages of adopting a more enquiring approach, involving a greater use of discussion, is that pupils tend to be forthcoming with ideas – both fruitful and unfruitful. Dealing with good ideas is generally a fairly easy matter; the bad ones sometimes present problems, particularly that of not discouraging the pupils who have them. Adopting the attitude that every idea is a good one just because it is an idea seems to us to have much to commend it at school level.

The value of discussion

The value of discussion in conveying the spirit and methods of scientific enquiry has already been mentioned. But it can also serve other purposes. It can provide an excellent opportunity for consolidation and revision. It can also enable students to work out hypotheses for themselves through a mutual exchange of ideas under the guidance of the teacher. Perhaps the most essential prelude to a successful discussion is adequate preparation of the material to be discussed, both by the teacher and the pupils (perhaps in preparation). The teacher will need to have a broad idea of the path the discussion is likely to follow and to have formulated the sort of opening questions that are likely to set it off on the right track. Teachers should take care, however, lest too much advance preparation should cause a potentially fruitful discussion to develop into a 'set-piece' affair of limited educational value.

Class practical work and demonstration

In this course a distinction has been drawn between demonstration by the teacher and the practical work carried out by the pupils. In class practical work the accent is either on *using* information, techniques, and concepts or *working them out*. It is essentially an investigatory or problem-solving activity.

A demonstration can also serve this purpose to some extent, particularly if the level of sophistication of the subject matter is relatively high. For instance, a demonstration by the teacher, aided by pupils' comments, may well give a better understanding of the controls in a particular experiment than if the pupils were left to their own devices.

In the main, however, demonstration is a method of imparting a piece of information, describing a concept or technique, or introducing a point for discussion, without the confusion arising from trial and error that invariably plays a part in class work. A topic demonstrated may, or may not, be something the pupils could have investigated themselves.

The evidence available suggests that for conveying information and describing concepts and techniques, a demonstration is as efficient as, if not better than, class work. Furthermore it is much quicker and, if used judiciously, can gain time in the course for more class practical work.

There is a place in biology teaching for both demonstration and class work. They are complementary to each other.

Group practical work

For the major part of their class work, pupils will be working in small groups, all performing the same investigation. But on occasions the work can be so arranged that each group is concerned with a particular different task; for instance when we wish to show that a particular concept applies to a whole range of organisms or when pupils are involved in a variety of different problems or investigations. The organization of such practical work differs little from normal routine, except that when the work is completed, it is important that the class as a whole should be made aware of the findings of the different groups. In addition, it is a good idea to depute a member of each group to give an account of the group's work. The results of each group investigation should be summarized on the board so that the whole class can make deductions from them.

Even when groups of pupils are all undertaking the same investigation it is wise to put a summary of the results on the board and to hold a discussion. In this way many important points, such as the purpose of controls, experimental error, and the making of hypotheses can be introduced. Class average results are far more significant than those obtained by single groups.

Long-term investigations

Some of the investigations undertaken during this course last two or three weeks or even longer. They need not occupy the whole time, however, and the art of using them for teaching is to integrate them with other work. To obtain the best value from such exercises, students should feel that they are individually responsible for their investigations throughout the whole period of their existence. This makes it necessary to allow time during class periods to make records and carry out any procedures needed to keep the experiments going. The natural tendency is to forget them once they have been set up, or to regard them as something of secondary importance, peripheral to the usual short-term class work. Here again, discussion will play an important part at the conclusion of the work so as to ensure that the main principles and implications of the investigation are not obscured by minor problems in experimental technique or other irrelevant difficulties.

Writing up practical work

There are many different ways of writing up practical work and all of them have their merits. It would be wrong to assert dogmatically that any one is the best. The following suggestions are intended to provide a rough guide which can be adapted to suit individual preferences.

a When writing up an investigation a pupil should aim to give in his own words a concise account of everything he did and why he did it. The use of distinct headings has much to commend it, particularly for juniors. These could be – problem, hypothesis, experimental method, results, deductions, suggestions for further investigations.

b Some experiments involve relatively complicated procedures. We find that pupils can benefit in their understanding if they are required to summarize the method in a brief statement or 'outline' – which may be only a few sentences long – before becoming involved in technical details which can obscure the overall pattern of procedure.

c It is not always necessary to write up each piece of work separately and completely. Where appropriate, summaries of a group of related experiments can be made.

d With reasonable guidance, pupils should become capable of writing up their practical work unaided. Dictated notes will seldom be needed except possibly to supplement the instructions for experimental procedure included in the *Text*.

e Pupils will find it helpful to cross-reference their own notebooks to the appropriate portions of the *Text*, particularly where practical work is involved. Such a practice greatly simplifies the problems of revision.

f The role of drawings and diagrams in biology can be greatly over-emphasized. We suggest that a drawing is only needed (*1*) to explain a point which cannot be adequately and concisely described in words, and (*2*) when a matter of structural detail either in an organism or a piece of apparatus needs careful exposition. In such instances, a large, clear drawing can be a valuable aid to understanding. Small sketches and simple schematic diagrams should also be encouraged but only where they clarify a situation which is inadequately covered by a verbal description.

Teaching elementary fieldwork

a Is fieldwork really necessary?
We hear many statements these days attempting to justify the inclusion of fieldwork in an elementary course: that it promotes a love of nature; that it enables children to see and handle living things; that it develops powers of observation; that it satisfies an acquisitive instinct; and so forth. Such arguments surely overlook the main issue. Among the important general principles in biology are those concerning the behaviour of *populations* and the relationships of the plants and animals which compose them. The principles of spatial distribution, succession, competition, and food-chains, to mention just a few, are fundamental to our understanding of the subject at any level. Moreover, they are aspects which, in the last resort, can only be studied satisfactorily in the field. Either we must include them, and in so doing concede the importance of fieldwork; or we must disregard them altogether. There seem to us to be no half measures; and for this reason, we have adopted the first policy unreservedly.

b The selection of organisms and sites for study
One of the superstitions that has adversely affected the teaching of fieldwork in schools is the belief that it is necessary to study a 'well-defined habitat' – whatever this may mean. Unfortunately, such an idea has been perpetuated in some examinations. All too frequently, the result is a decorative map and a long list of names, neither of which is likely to contribute much of value to biological understanding.

The present course in ecology attempts to break new ground in two respects:

1 It is so closely integrated with the relevant work in physiology and behaviour that it is sometimes difficult to discern where one ends and the other begins.

2 It is concerned with illustrating certain clear-cut ecological principles.

In designing our plan of work, we have been acutely aware of the wide variety of circumstances for which we must provide, ranging from luxuriant conditions of the South to extreme industrialization in the Midlands and North. Our plan has therefore been to state clearly the principles we wish to illustrate, then to devise simple experiments which, we hope, the great majority of schools will be able to carry out, even under the most inimical conditions.

The *Guide* contains a number of suggested alternatives, both in the selection of organisms and methods of working, which might be used, should the ones given in the *Text* prove inappropriate. A resourceful teacher will devise alternatives of his own. For schools in country districts, there will clearly be a wide choice of materials, any of which might be considered. At the opposite end of the scale, it may well be that there are schools where the local conditions are such as to defeat the ingenuity of even the most adaptable teacher. The solution then would seem to be periodic planned excursions.

The flexibility of material, and the variability of local conditions, preclude the anticipation of specific, hard-and-fast results. We have a situation which is essentially one of research, the students and teacher working together as a team, pursuing a line of enquiry and maintaining an open-minded, critical approach. The fundamental point to bear in mind is that we are concerned neither with specific organisms nor with rigidly prescribed experiments, but with *principles*. How these principles are set forth to students must be left to the discretion of the teacher. From time to time, seasonal considerations necessitate breaks in the smooth continuity of the course. The very nature of fieldwork makes this unavoidable. The fact that such breaks occur has its value in training students to be aware of the part played by the seasonal rhythm in the lives of animals and plants.

We hope that a clear statement of aims, and suggestions for practical work, both in the *Guide* and the *Text*, may assist teachers to exploit their opportunities and to carry out careful planning in advance.

References and indexes

For ease of reference the *Guide* chapters bear the same main headings and numbers as those in the *Texts*. Separate indexes wil be found at the back of every *Text* and *Guide*.

Preface
to the second edition

The revision of the Nuffield O-level Biology *Texts* and *Guides* has come about as the result of an evaluation of the materials after they had been used in schools for a number of years. A large number of teachers who had experience of using the materials contributed to the evaluation by answering a questionnaire and some of the teachers and many of their pupils provided additional information during visits to schools. The decision to revise was a direct result of the response from the teachers, and their views and those of their pupils have always formed the basis for the revision.

The objectives of the course

The original aims of the course remain the same, but in an attempt to give greater emphasis to specific objectives these have been stated in the *Guides*, usually for each main section of a chapter, but sometimes for the whole chapter and occasionally for sub-sections.

These are not exclusive and there may be other objectives implicit in the work or ones which a teacher may wish to emphasize particularly.

The content of the course

The subject matter also remains fundamentally the same, but is arranged in four *Texts* with corresponding *Guides*.

The first *Text* forms an introductory course which covers approximately two years' work and the other three *Texts* represent about another three years' work between them. *Text* 1 will probably be used in the first two years of a five year course and *Text* 4 in the last; the material in *Texts* 2 and 3 may be introduced in whichever of the other two years the teacher wishes. The sequence of the work in the books should be decided by the teacher although some suggestions for possible routes through the books have been given.

Also, in the *Guides*, an outline of the work covered in each topic has been provided so that the teacher can easily see the topic as a whole and decide how to develop it. It may be a very good idea to make this outline apparent to the pupils.

Practical work in the course

There is still an emphasis on group practical work, but the problems of long-term practical work are recognized and some data have been provided in the *Guides* for teachers to use when investigations prove too difficult to complete in schools. Some of the experiments are complementary and every practical investigation is not always required for a point to be made satisfactorily.

Questions

Questions which the pupils are particularly expected to answer have been clearly distinguished in the *Texts* and are numbered. The questions are repeated in the *Guides* with answers or with comments on the answers to be expected. These answers or comments should only be treated as a general guide to what is required and they should not restrict pupils from exploring points further.

SI units have been substituted and certain biological ideas have been brought up to date. Background reading has been made as relevant as possible and there is greater emphasis on social aspects of biology in many instances.

Grace Monger, General Editor

The perpetuation of life

The perpetuation of life
Relationships between the different topics

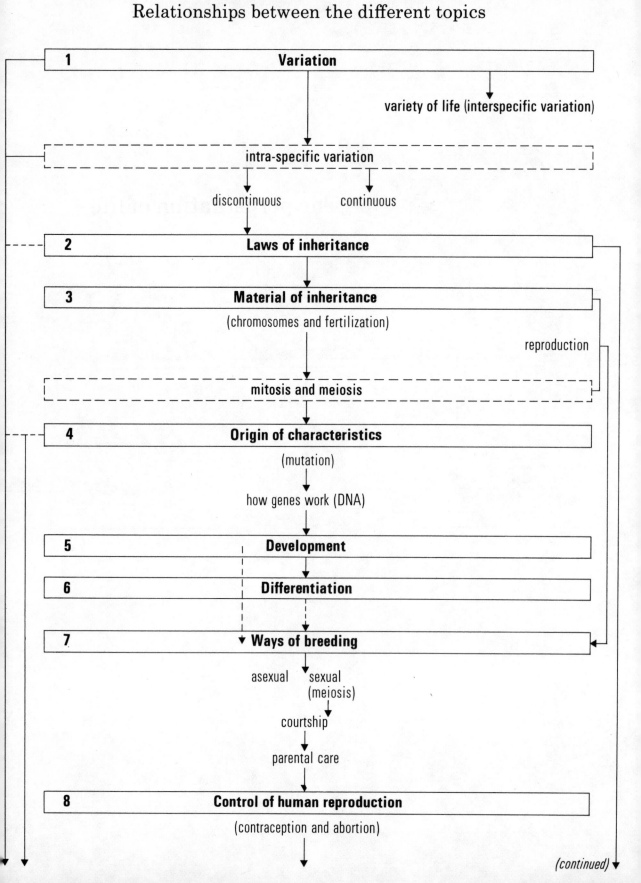

1   **Variation**

variety of life (interspecific variation)

intra-specific variation

discontinuous   continuous

2   **Laws of inheritance**

3   **Material of inheritance**

(chromosomes and fertilization)

reproduction

mitosis and meiosis

4   **Origin of characteristics**

(mutation)

how genes work (DNA)

5   **Development**

6   **Differentiation**

7   ▼ **Ways of breeding**

asexual   sexual
(meiosis)

courtship

parental care

8   **Control of human reproduction**

(contraception and abortion)

(continued)

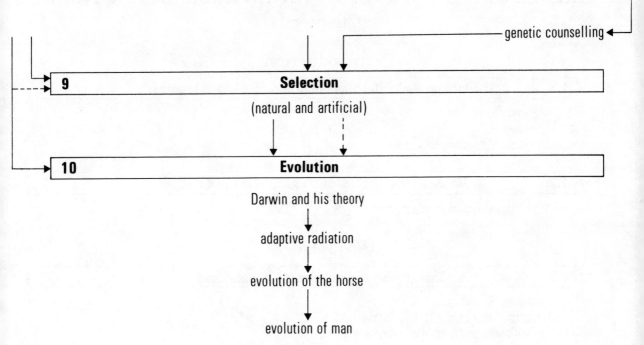

genetic counselling

9	**Selection**

(natural and artificial)

10	**Evolution**

Darwin and his theory

adaptive radiation

evolution of the horse

evolution of man

Outline of the work on variation and classification
(Chapter 1)

Introduction

1.1	Similarities and differences in grouping species together

Interspecific variation
Taxonomy
Artifical and natural classification

1.2	An international system of classification and naming

The binomial system of classification

Development

1.3	Similarities and differences in ourselves

Intra-specific variation
Eye colour, hair colour, height, mass
Discontinuous and continuous variation

1.4	Similarities and differences in blood groups

Discontinuous variation
Blood groups and transfusion

1.5	Immunity and transplants

Antigen–antibody reactions

1.6	Similarities and differences in finger prints

Intra-specific variation and systems of classification

Background reading	Racial conflict

Summary
to be made of the work on classification

1
Similarities and differences in living things

Characteristics are important in looking at living organisms. We can learn to look for important ones from amongst less relevant ones, depending on the purposes of the investigation. The study of similarities and differences falls broadly into two categories; those within a species (intra-specific variation), and those between different species (interspecific variation).

Introduction
(Sections 1.1 and 1.2)

This chapter looks at intra- and interspecific variation. The order suggested leaves the main material on intraspecific variation until the last parts of the chapter as it leads on to the material in later chapters. It would be perfectly possible to reverse this order or to leave the material on interspecific variation until evolution is dealt with.

What is a species? Discussion of this important topic is probably best left until the work on genetics and breeding has been covered. However, it is important at this stage to have some idea of what a species is and of how it is of central importance in thinking about biological material.

The way the definition is expanded will depend mainly on the way the word 'normally' is used with reference to inter-breeding. If two groups of organisms – populations – are separated by high mountains and only inter-breed in captivity, can they be considered as species? Another criterion of a species is that all its members must produce viable offspring that are not at a disadvantage compared with the parents. The mule, lion–tiger hybrids, numerous wildfowl (such as the Slimbridge pen) are evidence of the blurred nature of any definition of a species. Plants that only reproduce asexually do not fit into any definition which includes sexual reproduction as an important criterion.

There is considerable room for subjective judgments but the only test is whether the definition is useful in practice. This means that different definitions are going to be used by different people for different purposes.

These are some of the considerations that will emerge from work in later chapters and it will probably be best to leave a thorough discussion until later if it is attempted at all.

Q1 Can you remember approximately how many species of animals and plants you have found in any habitat that you have studied?

The actual number of species will vary according to the habitat but the important thing is the variety of species present in a habitat.

1.1 Similarities and differences in grouping species together

Objectives

1 to examine the problem of selecting characteristics for the purpose of classification

2 to show that there are two kinds of classification – artificial and natural

Apparatus and materials
Per pair:
4–5 whole plants in flower, each of different species
hand lens
razor blade
pair of forceps

Selection of plant material
It is possible to use a wide range of plants for this work. They can be locally common wild plants or cultivated plants, although the use of the latter would complicate the use of the flora. It is important to restrict the range of plants used to four or five species. At least two of the species should be in the same genus. The other two or three should perhaps be in the same family. Plants from the families Labiatae, Papilionaceae, Geraniaceae, and Compositae would be particularly suitable.

The illustrations in figure 2 of the *Text* would enable a few characteristics to be looked at but should form an adjunct to the work of the section rather than a substitute.

The selection of characteristics for comparison
It is important to develop the idea that some characteristics are important and others are not. What is needed is characteristics that are significant. This is a matter for class discussion. Below are some possible features that might be used.

a Characteristics of the flower	b Characteristics of the vegetative parts of the plant
1 How many petals? 1, 2, 3, 4, 5, 6, 7–10, many?	*1* Leaf shapes
2 How many sepals?	*2* Leaf arrangements
3 How many stamens?	*3* Growth habit (trees, shrubs, herbs)
4 The nature of the ovary and the stigma	*4* Hairs and spines etc.
5 The arrangement of the flowers into heads	*5* Bulbs, corms, tubers, bulbils, runners, etc.
6 Are there any other obvious features?	*6* Are there any other obvious features?

This would be a good opportunity to revise plant structure.

Recording the distribution of characteristics
The student should use table 1 in the *Text* as the basis for his recordings.

If this work is being done at a time when flowers are not available, any collections of preserved material could be used. (See 'Specimens, preservation of' in Appendix 2, page 221.)

The bird watchers could use locally common species, such as the tits, gulls, ducks, or finches. With these groups, characteristics of behaviour and feeding could also be used.

It would also be possible to use various wallcharts such as those produced by the Royal Society for the Protection of Birds. Emphasis should be placed on the selection of significant characteristics for comparison. Standard works on the identification of British birds, for example, *Field guide to the birds of Britain and Europe* by Peterson, Mountford, and Hollom can be used for taxonomic reference. (See Reference material on page 19).

1.2 An international system of classification and naming

Objective

To show that the wide variety of living things needs organizing for proper study

Q1 What are the Latin names for the flowers that you were studying in section 1.1?

Q2 How many of them have the same generic name?

These questions will only be relevant to the preceding investigation if the plants used have been from the same genus or if some of them have the same specific names. It would perhaps be better to choose examples specifically to illustrate this point, which is posed in question 4.

Q3 How many of them have the same specific name?

Q4 What is the significance of your answers to questions 1 to 3?

The organisms with the same generic name have many characteristics in common, that is, many similarities, while those in different genera have fewer similarities and more differences. In evolutionary terms they are thought to be closely related. This is not the best place to introduce evolutionary ideas but it forms a link with the work on evolution in the last chapter.

Students could be asked to fill in the table for man in the same way as it has been filled in for the lion. The following question could be useful.

'At which level of grouping do lions and men belong to the same category? What does this suggest about the number of similarities and differences between these two species?'

The answer is 'at the level of the class'. This means that there are relatively few similarities. These similarities are the characteristics of mammals, such as warm blood, suckling of young, complex dentition, hair, and a number of skeletal features.

There are a number of categories intermediate to those given for the lion but these do not seem important at this stage. They are the sub-family, Felinae; sub-order, Fissipedia; and sub-phylum, Vertebrata.

Development
(Section 1.3 to end of chapter)

1.3 Similarities and differences in ourselves

Objective

To show that, roughly speaking, variation can be continuous or discontinuous according to whether the differences are quantitative or qualitative

This introduces the idea of qualitative and quantitative differences with regard to characteristics. It also examines this variation as being continuous or discontinuous.

The techniques for collecting data of continuously varying kinds involve measurement and dividing the range of observations into arbitrarily chosen intervals. The data can then be plotted as a frequency histogram. In this section the data for height and mass are of this form. Data from discontinuously varying material do not demand such complicated methods of presentation. It is sufficient to count the number of individuals showing the particular characteristic and to record this on a bar chart. The data for hair colour, to some extent, and, rather better, the data on eye colour, are of this kind. Even clearer are the blood groups investigated in the next section.

The perpetuation of life

The statement that man is generally considered to be one species could lead into a brief consideration of what is meant by interspecific and intra-specific variation.

Q1 Is there any evidence that some characteristics do not fall into sharply defined categories?

Yes.

Q2 What characteristics of the people in your class or of others you have met are of this sort?

Very many features show variation of this sort. Certainly, height and mass do, also hair colour, particularly the brown hair characteristic. Eye colour may also show this variation, within the categories of brown and blue which are easily defined as discontinuously varying features.

Q3 Is there any evidence that some characteristics do not grade into others?

Yes.

Q4 What characteristics of this sort can you see in your class or have you met in others?

In the data on ourselves, eye colour is easily divided into brown (with pigment) and blue (without pigment), although the precise expression of these characteristics varies, in the case of brown from very light hazel to a very dark brown. In the case of blue, there are various shades of blue, grey, and violet. With hair colour the range can be from black to brown to blond but red is to some extent qualitatively different and can usually be separated.

Q5 To what extent is it difficult to say that characteristics show only continuous or only discontinuous variation?

This is an opportunity to bring out the indistinctness of the definitions when they are applied to living material. To a greater or lesser extent all characteristics show some discontinuous, and some continuous variation. The definitions are useful because they may point to the most useful way of collecting data.

Q6 How have the similarities and differences come about?

The question leads into the next chapters. The only answer required here is 'by inheritance', although later, mutation and selection would also be appropriate suggestions.

Q7 What sort of characteristics would you choose to study in investigating the answer to question 6?

The use of characteristics that are discontinuously variable would greatly help investigation, as we would be dealing with simple presence–absence information. It is true, however, that this may severely limit the conclusions and generalizations that can be made. It is a basic tenet of experimental science that the simplest investigations are made first but that caution must be used in making generalizations.

1.4 Similarities and differences in blood groups

Objective

To study a system of classification based on distinct categories, that is, on discontinuous variation

The A–B–O blood groups are good examples of characteristics that can be assessed objectively. With rare exceptions they are easily distinguished. They provide a good illustration of discontinuous variation.

In contrast, finger prints, which are the subject of 1.6, cannot be assessed as objectively. They vary from one individual to another and some of their characteristics show continuous variation.

The practical work of sections 1.1 and 1.3 is important mainly as providing exercises in classification rather than as a means of obtaining specific information. The students should be able to apply their understanding of the problems of classification to other organisms and other characteristics than those dealt with here.

During the practical work, discussion with groups of students can help to emphasize:

a That there is a need for careful identification.
b That distinctions between organisms based on particular characteristics do not necessarily apply to all their characteristics.

Table 1 indicates when transfusion is possible between blood groups. It can be used to check the answers students give to question 4.

Substances (antibodies) in receiver's serum	Substances (antigens) in donor's red cells			
	A (group A)	B (group B)	AB (group AB)	NONE (group O)
anti-B (group A)	TP	TNP	TNP	TP
anti-A (group B)	TNP	TP	TNP	TP
None (group AB)	TP	TP	TP	TP
anti-A and anti-B (group O)	TNP	TNP	TNP	TP

Table 1
TP = transfusion possible
TNP = transfusion not possible

Human blood groups

Human red cells may contain many different blood groups – red cell antigens or agglutinogens – all of which are inherited characteristics. The serum of a person's blood may contain antibodies or agglutinins which may react with the red cells of some other people if mixed with them. The reaction results in the 'clumping' or agglutination of the red cells – a process not to be confused with clotting. The antibodies may occur *naturally* or may result from the transfusion of blood from another person. (Antibodies of the second kind are called immune.)

Because antibodies react with the red cells of other people, the blood groups of patients in hospitals have to be determined. Then they can be given blood which matches their own. Otherwise a transfusion could kill them. The major blood groups were the first to be found (Landsteiner, 1900), and are known as the A–B–O groups. The antibodies to the A and B red-cell antigens occur naturally – and are called anti-A and anti-B. This is Landsteiner's Law, and is the basis of classifying all humans into four groups by using anti-A and anti-B. (Since the classification involves testing the blood, the term 'blood group' has arisen.)

The study of blood groups in man can be used as a starting point for more advanced studies, such as genetics (see, for example, Lawler and Lawler, 1953) and ethnology. Some of these studies could be undertaken by bright fifth-form students.

The frequencies of blood groups among racial groups differ from one group to another, and this is a means of establishing relationships between different racial groups – of following the movement of people by migration or invasion, and so on.

The data in *table 2*, coupled with geographical studies, suggest the following possible conclusions:
European gypsies may have originated in India – probably in the north-west region.
The early inhabitants of the British Isles were largely of group O, and their descendants retreated to the north and west before successive waves of invaders, such as the Anglo-Saxons entering from the south and east who possessed a much higher percentage of group A. This is thought to be the case.

Studies of this kind can usefully complement historical and geographical studies. The data in the table could be used to devise problems for students in follow-up work.

It is worth emphasizing that, except in very isolated populations such as some of the South American native tribes, many of which are exclusively of blood group O, all the blood groups are found in all populations. Populations are distinguished by differences in the frequency of the various blood groups.

| | **Blood group** | | | |
	O	A	B	AB
English	47	42	8	3
Welsh	49	38	10	3
Scots	52	34	11	3
Irish	54	32	11	3
Gypsies in Europe	31	27	35	7
Indians	33	24	34	9

Table 2
Percentage of persons from different locations with different blood groups
After Mourant, A. E. (1962), Blood groups and the study of mankind, *Ministry of Health.*

Students might consider other implications of blood grouping:

c *Anthropology and evolution* Since some of the blood groups occurring in man also occur in higher apes there is evidence of a close evolutionary connection.

d *Criminology* The identification of the blood groups of blood stains associated with a crime sometimes provides vital evidence in criminal investigations.

e *Human medicine* The role of blood grouping in blood transfusion could lead to a fuller consideration of the National Blood Transfusion Service. The role of the Rhesus blood groups in pregnancy could also be dealt with.

The literature cited at the end of this chapter will provide material for this follow-up work.

Obtaining blood samples

Apparatus and materials
Per pair:
2 clean glass slides per blood sample to be taken
wax pencil or glass writer
2 clean droppers or pipettes
clean glass rod

Per pupil:
sterile lancet (there are several types, available from biological suppliers)

Access to:
anti-A serum
anti-B serum
saline solution; this may not be necessary

Reagents
a A drop of physiological saline solution (0.85% sodium chloride) should be added if necessary, to slow down the rate at which the blood dries up.

b *Anti-A and anti-B grouping sera* These *may* be obtainable from the nearest regional laboratory of the National Blood Transfusion Service, the address of which is often displayed in GPO branch offices. The address may also be obtained at local hospitals.
Eldon ABO Rh⁺ cards These are cards with four spots on them, each impregnated with single anti-serum. A drop of blood is put on each spot and according to whether clumping occurs or not the blood groups of both the A–B–O and Rh types can be identified.
The cards are simple and foolproof in use. However, they can be used once only and are, therefore, expensive.

The Department of Education and Science recommends

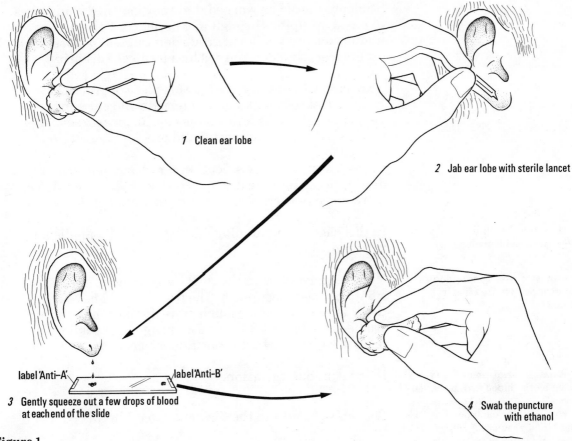

1 Clean ear lobe

2 Jab ear lobe with sterile lancet

label 'Anti–A'   label 'Anti–B'

3 Gently squeeze out a few drops of blood
at each end of the slide

4 Swab the puncture
with ethanol

Figure 1
Taking a blood sample.

that only staff should obtain blood from students. The
procedure is as follows (see *figure 1*).

1   Obtain blood from an ear lobe. Before puncturing, wash the
part with soap and water. Rub dry. Then rub down with
ethanol to remove grease and rub dry again. If desired,
swab with ether to anaesthetize the area.

2   Use a sterile lancet. Do not remove it from the sterile packet
until it is time to use it.
A quick, sharp jab through the skin is needed. Do not put
the lancet on the skin and press down. Literally only a
pin-prick is needed.
**Sterile lancets must be used only once and then
discarded.**
Unless this rule is strictly observed, there is a serious risk
of infection. Some viruses which affect the blood are
unaffected by alcohol and other chemical disinfectants.

3   Squeeze the pricked area and drop blood on two clean glass
slides, or at either end of one.

4   Swab the punctured area with ethanol. This will help to
stop bleeding. Cover with a small dressing if the area is
liable to come quickly into contact with dirty materials.
Otherwise, exposure to air is good for healing.

Students should be warned that they may make mistakes in blood grouping through inexperience. This will prevent undue concern if a blood group determined in class work differs from that determined by a medical authority.

Two exercises are involved in this piece of practical work.
1 Using a test to detect a characteristic and carrying it out very carefully. This follows the careful measurement needed in 1.3 and the careful and precise observation needed in 1.1.
2 Using a system of classification for convenience of description. Here the use in medicine is clear and this is introduced in the reading at the beginning of the section.

Q1 How can the similarities and differences be identified?

In this case an immunological test, carefully applied, is required.

Q2 Can we classify individuals into groups that are clearly defined?

Yes. Intermediates in the A–B–O system are very rare and would not be detectable by the test that will have been used. These groups are much more distinct than those studied in section 1.3 and are a very good example of a feature which shows discontinuous variation.

Q3 What is variation of the type shown by the blood groups called?

Discontinuous variation.

Q4 Work out for each person what blood group or groups could be used if he or she required a transfusion. Remember that it is the reaction of the donated red cells with the serum of the recipient that is important.

See tables 3 and 4 in the *Text* and *table 1* in this *Guide*.

Q5 What are the proportions of the different groups in the class?

See *table 2* in this *Guide*. This may be a good point to discuss the ethnological evidence that the proportions give.

Q6 You might obtain data on the distribution of blood groups in different races and compare them with the result you got for your class.

1.5 Immunity and transplants

This is a conceptually difficult section that is, however, of immediate relevance to everybody. The ideas of what antibodies are have already been investigated in a practical manner in the previous section. The material has been laid out more or less as an exercise in comprehension which might be attempted for preparation and then form the basis of a theory lesson. The ethics of transplants could also be brought in at this point and comparison of organ transplants with blood transfusion may be useful. The transience of red blood cells in relation to the ultimate aim of organ transplants is a good point of discussion which also provides useful revision of the function of the blood.

Avoid saying that changes in influenza virus are mutations; nevertheless, you could bring this up when mutation is being considered.

Q1 What is immunity?

A definition should include 'resistance to disease'.

Q2 How can immunity be built up by an organism?

Naturally, by getting the disease and the build-up of antibodies during recovery. Artificially, by the injection of dead or weakened pathogens.

Q3 What is an antibody?

Simple definitions are expected for questions 3 and 4.

Q4 What is an antigen?

Q5 What reasons are there for immunity not always being complete?

1 The production of antibodies by the body falls off with time, and, therefore, immunity wears off.
2 There may be changes in the pathogens and, since the antibody is specific to an antigen, a new antibody is required.

1.6 Similarities and differences in finger prints

Objective

To examine and assess the problems and complexities of an important system of classification based partly on discontinuous, and partly on continuous variation

Apparatus and materials
Per pair:
several sheets of white paper
access to 70% ethanol
cottonwool or cloths for cleaning fingers with ethanol
glue *or*
adhesive tape for fixing the prints into a collection

Per 3 or 4 groups:
office stamp pad

Note: Schools are advised to use an office stamp pad because this avoids the mess involved in the best way of taking prints. If, however, the teacher wants students to attempt the best way, they should use fingerprint ink and a brass plate or glass plate or even a Petri dish top. Fingerprint ink can be obtained from Heath, Hicks & Perken, 8 Hatton Garden, London EC1. It may also be obtainable at shops selling artists' materials.

This is one opportunity for showing that each individual is unique. In fact, each individual finger has a unique pattern of ridges. It is preferable not to deal at this stage with the uniqueness of sexually produced organisms and the relevance of this to evolution and selection. Nevertheless, the work of this section will form a link with later chapters when these subjects are being covered.

The framework for the investigation is the use of finger prints for crime detection and the *two* assumptions that this rests on.

Evidence for the first of these assumptions, namely the constancy of finger prints with time, is contained in a not very clear pair of prints taken at an interval of 54 years. The intention is to show the technique that can be used rather than conclusive proof. The prints shown were those of Sir William Herschel (1833–1917) who, with Sir Francis Galton and Sir Edward Henry, and others, evolved the system of classifying finger prints called the Henry system that is the basis of those used by police forces throughout the world.

Obtaining finger prints

If the class is using fingerprint ink, spread a very thin layer on a flat glass plate with a roller, as a means of applying the ink to the finger. But it is simpler to use an office stamp pad.

A collection of finger prints

These can be cut out and stuck into an exercise book with separate pages devoted to each of the categories. Alternatively, a folder of plain A4 sheets of paper can be used. Each sheet can hold about 20 prints. There are eight categories that are easily distinguished and one of these – ulnar loops – will contribute 60 to 70 per cent of the prints and will require two sheets. The rest of the categories should fit onto a half sheet of A4 paper. This assumes that only about 50 prints are collected. It is important for each print to be labelled with the name of the person and which finger of which hand it is from, as well as with the group it belongs to. The fingers can be abbreviated as Th (thumb), I (index), M (middle), R (ring), and L (little). (See *figure 2*.)

Q1 What are the characteristics of the prints? Can you detect the types shown in *figure 9*?

Each print shows a pattern of ridges. There are two types of arrangement of these ridges which can act as fixed points. The core is the centre of a loop or whorl. A triradius is the point from which three lines of ridges come – see figure 9 of the *Text*. You can detect the types shown in that figure. It may be possible to find one or two others, but see below.

Q2 Are the prints from different fingers of one person the same or different?

Different.

Name M. R. Smith (this information needs to be
Hand L.H. recorded next to *each* print)
Finger I
Group A

Figure 2
Collecting and recording
fingerprints.

Q3 Do all prints fit into distinct groups as shown in *figure 9*?

Yes. See the answer to question 5. It is an example of discontinuous variation in that there are very few intermediates.

Q4 What are the proportions of each type you have found in your sample?

Approximately, loops, 70%; twinned loops, 13%; whorls, 13%; arches, 4%; and composites, rare.

Q5 Is it possible for any of these groups to be subdivided?

Yes. Arches can be divided into simple arches (A), in which there is no triradius, and tented arches (T), in which the centre of the pattern is held up by a vertical ridge which originates from or near the triradius. Loops can be divided into ulnar loops (UL) and radial loops (RL) according to which side of the finger the loop opens out on. The print of an ulnar loop from the left hand opens to the left as you look at it. A radial loop from the left hand opens to the right as you look at it. The loops on the right hand are the opposite. The anatomical definition is based on the side on which the loop opens: the thumb side of the hand is the radial side. Twinned loops (TL) possess two distinct loops and one is the ascending loop and the other is the descending loop. There are two triradii which must be on opposite sides of the ascending loop.
If the two triradii are on the same side of the ascending loop then it is called a lateral pocket loop (LP). This category is rare and it may be simpler to treat this group as being a composite. Composites usually have more than two triradii.

Q6 Might it be easier, in order to subdivide the groups for the purpose of classification, if you could obtain better quality prints than you are likely to have done? What features would you like to see more clearly? At New Scotland Yard the prints are classified so that the print of an unknown person can easily be checked.

Subdivision of the prints for the purpose of identification involves characteristics of the prints that will not be clear on impressions of this sort. These are:
1 Ridge counts between core and triradius.
2 The presence and position of bifurcations, ridge terminations, islands, and independent ridges. At least 16 of these characteristics must be the same before two prints are declared the same.

---

3 A number of other features could be used in school but have no wide general use. It is not possible to use measurement between fixed points as an accurate guide as this will vary according to the pressure applied.

Q7 How would you devise a scheme of classification for your collection by which you could check a person's print quickly and accurately against those you have obtained?

Part of the answer is implicit in the division and further subdivision of prints into groups and sub-groups. See the answer to question 6.

Q8 Is the scheme of classification that you have devised an example of a *natural classification* or of an *artificial classification*?

It is an artificial one, as it only deals with one characteristic at a time, rather than a number of characteristics at each stage.

Q9 Do your observations support the assumption, mentioned earlier in the section, that no two people have identical prints?

Yes.

Background reading

Q1 What were the features or characteristics that the Nazi regime used to identify the Aryan race?

Racial conflict

Being slim, athletic, tall, long-headed, blue-eyed, fair-haired, etc.

Q2 How accurate or satisfactory, as guides to identification, were these features in practice? Give examples to support your view.

For discussion. Students will soon see the difficulties in relation to the appearance of the Nazi leaders, and their allies the Italians and Japanese.
Blood groups and race could be a convenient topic here.

Students are asked to look at the data on some of the features of the different races of mankind which are given in table 6 of the *Text*.

Q3 Can you classify the races into groups or sub-groups? Show your scheme of classification by joining the numbers (1—8) by lines or brackets.

This will not be a success unless students have been given some idea of *how* to display their schemes.

For example:

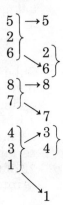

Summary

At this point students should be asked to make their own summary of the work of this chapter.

Reference material

Books

*reading suitable for students

Bibby, C. (1959) *Race, prejudice and education.* Heinemann.
Boorman, K., and Dodd, B. (1970) *Introduction to blood group serology.* Churchill.
Brierley, J. (1973) *The thinking machine.* Heinemann.
*Cherrill, F. R. (1954) *The fingerprint system of Scotland Yard.* H.M.S.O.
Clapham, A. R., Tutin, T. G., and Warburg, E. F. (1962) *Flora of the British Isles.* Cambridge University Press.
Elton, C. S. (1966) *The pattern of animal communities.* Methuen.
Harrison, J. Heslop (1953) *New concepts in flowering-plant taxonomy.* Heinemann. (A concise account of present-day outlooks and methods in taxonomy.)
Martin, W. Keble (1969) *The concise British flora in colour.* Ebury Press and Michael Joseph. Available also in a paperback edition (1972). Sphere Books.
Lawler, S. D., and Lawler, L. J. (1953) *Human blood groups and inheritance.* 2nd edition. Heinemann.
*McClintock, D., and Fitter, R. S. R. (1956) *Pocket guide to wild flowers.* Collins.
*Peterson, R., Mountford, G., and Hollom, P. A. D. (1965) *Field guide to the birds of Britain and Europe.* Collins.
United Nations Educational Scientific and Cultural Organization (1952) *What is race?* UNESCO, 19 Avenue Kleber, Paris.
*Various authors *The race question in modern science.* UNESCO. Available from H.M.S.O. (Covers the subject from many aspects; political, psychological, historical, sociological, biological, etc. The individual articles are also sold separately. 'Race and biology' by L. C. Dunn can be recommended.)

Articles

David, T. J. (1971) 'Dermatoglyphics in medicine'. *Bristol medico-chirurgical journal,* Vol. **86**, pages 19–26.

Wallcharts

Royal Society for the Protection of Birds: wallcharts of British birds. R.S.P.B., The Lodge, Sandy, Bedfordshire SG19 2DL.

Pamphlets

Various leaflets are available from the National Blood Transfusion Service. Some of these are very suitable for students to read. Teachers should write to the Regional Director for copies.

Outline of the work on inheritance
(Chapters 2 and 3)

Introduction

2.1	**Where do living things come from?**

Spontaneous generation or reproduction?

Development

2.2	**Are the characteristics of an organism inherited?**

The Wedgwood–Darwin–Galton family tree

2.3	**How are characteristics inherited?**

Breeding with *Drosophila*

2.4	**Using models**

Practical investigation, using beads

2.5	**Investigations with other organisms**

Inheritance in *Drosophila* and other organisms

2.6	**Inheritance in seedlings**

Background reading	**Spontaneous generation**

3.1	**A living thing begins**

Fertilization in *Pomatoceros*

3.2	**A model of inheritance**

$$2n \nearrow^{n} \searrow 2n$$
$$\searrow_{n} \nearrow$$

3.3	**Cell division in growing tissues**

3.4	**Cell division in making gametes**

3.5 **Comparison with the results of the breeding investigations of Chapter 2**

Summary

Background reading	**Is seeing believing?**

How do similarities and differences come about?

1 to make a distinction between reproduction and the concept of the spontaneous generation of organisms from non-living material

2 to introduce the idea that characteristics can be inherited through reproduction, but that the environment may also be influential

3 to study monohybrid inheritance through breeding investigations with *Drosophila* and other organisms

4 to establish that inherited characteristics are transmitted through 'factors' present as single entities in the gametes and as double entities in the zygotes

5 to devise a model of inheritance by using beads, to help to explain the results of the breeding investigations

6 to introduce the concept of probability

Sequence of practical work in Chapters 2 and 3

It is suggested that teachers tackle the work of Chapter 3 while the F_1 and F_2 generations of *Drosophila* are developing. A suitable sequence would be:

		Section
1	Demonstrate handling *Drosophila*	2.3
	Set up *Drosophila* cross	2.3
	Set up any long-term investigation	2.5
2	Construct bead model of F_1 and F_2	2.4

From this the model describing the 'pattern of inheritance' can be devised:

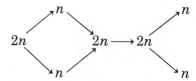

adults → gametes → zygote → adult → gametes

This logically turns our attention to zygote formation (fertilization) and cell division in the formation of a $2n$ adult from a $2n$ zygote (mitosis) and in the formation of n gametes from cells in a $2n$ adult (meiosis).

		Section
3	Fertilization in *Pomatoceros*	3.1
4	Score and analyse *Drosophila* F_1 (if you wish predict F_2 flies from the model)	2.3
5	Cell division in growing tissues	3.3
6	Cell division in making gametes	3.4
7	Score and analyse *Drosophila* F_2	2.3
8	Inheritance in seedlings (parents, F_1 and F_2)	2.6
9	Relate all the work to the model of inheritance $\begin{cases} 2.4 \\ 3.2 \\ 3.5 \end{cases}$	

The following is an *alternative* teaching sequence which takes Chapter 2 first and makes *Drosophila* optional:

		Section
1	Discuss *Drosophila* work or provide class with living or preserved parents, F_1 and F_2 to score and analyse	2.3
	Set up any long-term investigation	2.5
2	Inheritance in seedlings (parents, F_1 and F_2)	2.6
3	Bead model of inheritance	2.4

---

4   Relate items 2 and 3 and now move to Chapter 3
to study gametes and cell division as the
physical structures involved in inheritance

Before starting the work of this chapter, you might
present the students with a large diagram (on the
blackboard or using an overhead projection transparency)
to show the route you intend to take through these *two*
chapters.

Advanced preparation

(See Appendix 2, page 200, for *Drosophila* culture.)

To ensure that adult flies are available in sufficient
numbers on the correct day, one must follow a definite
routine, such as the following.

Day − 14   Set up cultures of normal, grey body (+) and
            ebony body (e) flies, preferably in duplicate.
            Incubate at 25 to 26 °C.

Day − 10   Larvae should be present. Remove parents.

Day − 5   Make up food and prepare tubes for crosses
            (20 for a class of 30) and for storing segregated
            flies.

Day − 4   Flies start to emerge. Segregate males and
            females from cultures cleared a.m. and p.m.
            To set up 20 crosses (10 of + female × e male
            and 10 of e female × + male), the minimum
            requirements will be 20 virgin + females and
            20 virgin e females, 30 + males and 30 e males,
            to allow crosses between two virgin females
            and three males in each tube.

Day 1   Make up crosses (also sub-cultures for future
            use).

Day + 4   ⎱ Class work begins. Class members tip out
or 5   ⎰ parents. Kill and examine them, checking
            characteristics of sex and body colour. The
            class should have been shown the
            characteristics of *Drosophila* and instructed
            in the experimental procedure in a previous
            lesson.

Day + 10 to ⎱ F_1 flies emerge. Class members make crosses,
Day + 14   ⎰ checking characteristics of the F_1 adults.

Day + 15 to ⎱ Larvae should be present in the tubes with
Day + 19   ⎰ F_1 crosses. Remove parents.

Day + 20 to ⎱ F_2 flies emerge. Remove and score for sex and
Day + 25   ⎰ body colour. Keep flies at room temperature
            if required later.

Introduction
(Section 2.1)

2.1 Where do living things come from?

Objective

To make a distinction between reproduction and the concept of the spontaneous generation of organisms from non-living material

The main topic of this chapter is inheritance, but it is as well to make students aware that the origin of organisms themselves is fundamental to the idea of the transmission of characteristics.

The idea of spontaneous generation is outlined in this section. The investigation with worms should help students to appreciate that the views on spontaneous generation held in past centuries could be plausible if they were not subjected to experiment.

With normal garden soil many specimens of the nematode *Rhabditis* will collect under pieces of putrefying earthworm. They are white, about 3 mm long, and can be seen with a hand lens. (See also section 6.2, page 106, of this *Guide*, on *Rhabditis* and cleavage.)

Q1 Explain what you find. Was Kircher's logic so ridiculous?

Students can compare this experiment with that of Kircher and be asked if they consider they have demonstrated an example of spontaneous generation. They will need to devise a control experiment to check their hypothesis. Sterilized soil is the obvious control to use.

Section 2.1 also provides a lead into the Background reading 'Spontaneous generation'. This illustrates the nature and importance of an experimental approach to biological problems and the way in which preconceptions can influence decisions.

It is suggested that the practical work of section 2.1 should be done by a small group of interested students as an extra-curricular project and not by the whole class. The findings can be discussed by the whole class at a time that does not unduly interrupt the work on inheritance.

Development
(Section 2.2 to end of Chapter 3)

2.2 Are the characteristics of an organism inherited?

Objective

To introduce the idea that characteristics can be inherited through reproduction, but that the environment may also be influential

Q1 Do you think that this pedigree shows that scientific ability is inherited?

Q2 What are your reasons?

Q3 What possible errors may there be in your deductions?

The main subject of the exercise on the Wedgwood–Darwin–Galton family tree is that the students should conclude, in their answers to questions 1 and 2, that the presence of a characteristic in several successive generations is an indication that it is inherited, and, in their answer to question 3, that the cultural environment in which the families lived probably contributed to the unusual extent to which their inherited characteristics could be expressed. One cannot conclude that people less fully exposed to such a cultural environment have less inherited scientific ability. The exercise also illustrates the difficulty of distinguishing between inherited and environmental aspects of human characteristics, and suggests that the small number of individuals in the investigation is not enough to justify firm conclusions. Thus the students are made aware that environmental influences can play a part in the development of characteristics. This point is returned to in Chapter 5 but it is important that it should also be borne in mind in the earlier work.

The question of the size of a sample is taken up in sections 2.3 and 2.4.

2.3 How are characteristics inherited?

Objectives

1 to study monohybrid inheritance through breeding investigations with *Drosophila*

2 to establish that inherited characteristics are transmitted through 'factors' present as single entities in the gametes and as double entities in the zygotes

Apparatus and materials
Per pair of students or for a demonstration:
Drosophila cultures for whole class
hand lens
white tile, or filter paper attached to the bench by adhesive tape
seeker, cardboard triangle, or small, finely pointed paint brush
bench lamp, unless there is very good laboratory lighting
labels
etherizer
emergency etherizer
small tube of 1:1 ether–ethanol and a pipette

Drosophila is suggested as suitable for this work because:

a It allows students to obtain realistic, firsthand data about inheritance through their own investigations in a relatively short period of time. Trials with fifth-year students have indicated that they get bored and confused if results are not obtained fairly quickly. It has been found that three weeks is the longest time suitable for an investigation through which a *new* concept is being taught. If possible, two weeks or less should be aimed at. While the class is waiting for the results, problems or discussions *relevant* to the investigations should, if possible, be introduced.

Once the students have understood a concept it is feasible to introduce long-term investigations lasting many weeks to confirm, or elaborate on, the short-term work. These are best carried out by individuals or small groups of students. For fuller comments on long-term investigations, refer to the Preface to the first edition.

b In the school trials in 1964 and 1965 it was found that on the whole, both fourth- and fifth-year students can successfully tackle practical work with *Drosophila*, and that they can be engrossed by it.

Either the students could read the first part of section 2.3 or there should be a short discussion based on it.

A demonstration of *Drosophila* technique could then be given. Techniques of culturing *Drosophila* are described in Appendix 2, page 200. A possible sequence is:

1 Show students in outline the preparation of the culture medium, sterilization, and incubation. (It is recommended that the students should not undertake these tasks in their class practical work. Some may do so at other times as assistants to the teacher or laboratory technician, or as part of project work.)

2 Demonstrate techniques of transferring and scoring flies.

3 Indicate to students the differences between male and female flies, and flies of normal body colour (grey) and ebony body. Prepared slides shown through a micro-projector, blackboard drawings, or enlarged photographs could be used. It could be pointed out here that scoring the flies is an exercise in distinguishing similarities and differences comparable to those undertaken in Chapter 1. Point out that newly emerged adults may not be very black in colour but ebony-bodied flies will show their colour within a day.

4 Show the film loop NBP–70 'Handling *Drosophila*'.

5 *Either* provide students with tubes containing the different types of flies and allow them (working in pairs and under your direct guidance) to perform the techniques in stages linked to the instructions in the *Text*, *or* proceed to class practical work.

The choice here depends on time and facilities. The parent flies that are scored in the first practical exercise are discarded, so errors of technique will not hinder the work. After the demonstration the film loop could be shown continuously for about ten minutes so that students can check on their own techniques as they try them out.

Class practical

It is best for students to work in pairs. It is important to aim at a brisk pace. With too much leisure in this type of work, some students tend to allow the incidental practicalities of the technique to become over-important, and the real point to be lost. Particular care also has to be taken not to let scoring flies occupy a disproportionate amount of time.

If the number of flies produced by the whole class is too great, samples can be scored. About 400 flies from each generation are quite sufficient. Distributed around a class, this means that each student can score, say, 12 to 20 flies. This can be done very quickly, possibly in about 10 minutes.

Discussion with students during the practical work is best confined to :
problems of technique
the outline and purpose of the work
what results the students expect to get, and their reasons.

The last topic should make it possible to anticipate difficulties in analysing the results. Students tend to have various preconceptions about inheritance which are frequently erroneous. These should be identified as soon as possible, and challenged with the appropriate results from the students' investigations.

Analysis of results and questions
The students should record their results in a table similar to table 7 of the *Text*. Class results should be displayed on the blackboard.

Between setting up the crosses and obtaining results, section 2.4 should be studied.

Q1 Is it correct to infer that the existence of pure line cultures, for different characteristics and containing hundreds of flies, is evidence that the characteristics are inherited?

As soon as the crosses have been set up, students should be invited to consider the first question.

Yes. Because flies of different characteristics are kept in identical conditions it is unlikely that particular characteristics can be determined by the environment. The large number of flies in a culture means that it is likely

The perpetuation of life

that all possible characteristics will be displayed by the flies. This argument is to be distinguished from that arising from the example of the Wedgwood–Darwin–Galton family tree, which is analogous to considering a pure line culture of flies without reference to cultures of flies with other characteristics.

q2 How can you explain the results obtained from investigations *b* and *c*?

None of the flies of the F_1 have ebony body, but this characteristic is present in the F_2. Thus, using the symbols + (= normal body) and e (= ebony body), the scheme shown in *figure 3* is likely to be produced.

Figure 3
A scheme showing the observed characteristics (phenotypes) of the fruit flies in each generation of the breeding experiment.

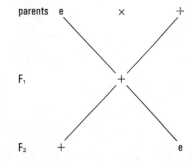

Able students may continue: 'Assuming that there is something in the flies responsible for the characteristics of grey and ebony body the disappearance and reappearance of the ebony body characteristic can be explained'.

To understand this, it must be assumed:
1 that the things responsible for the characteristics in the organism exist in pairs and singly in the gametes, *and*
2 that when the members of a pair are different only one can produce its characteristic. That is, one is dominant over the other.

If, as is likely, students find it difficult to work this out without assistance, they can be asked to consider what happens in reproduction and be referred to the model outlined in section 2.4. Suggest to the students that they try to explain the results of their breeding experiments with that in mind. The scheme in *figure 4* can then be worked out.

The terms homozygote and heterozygote can be introduced here but probably it is unnecessary.

The role of probability in achieving the results of the F_2 should be appreciated by the students. With each act of fertilization any one of the four possible combinations could theoretically be produced. However, given a large enough number of acts of fertilization (100 or more) the offspring are produced in approximate proportion to the

Figure 4
A model scheme showing the
distribution of the things (genes)
responsible for the characteristics
of grey body and ebony body in the
gametes and zygotes of each
generation of the breeding
experiment.

probability of each combination. This is $\frac{1}{4}$. In fact, three of
the combinations, $+\,+$, $+\,e$, and $+\,e$, produce flies of
normal body colour and one, ee, produces ebony-bodied
flies. Thus, the probability of obtaining flies of normal body
colour is $\frac{3}{4}$ and the probability of obtaining ebony-bodied
flies $\frac{1}{4}$. It is possible that the results of individual pairs of
students will not always show this, but the total class
results should.

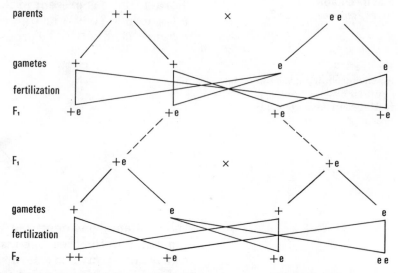

The analysis of the results could be undertaken completely
as a discussion. If the class has understood the work of
section 2.4, students will not have much difficulty in
working out the answers for themselves.

An alternative method of layout for this analysis is to
use squares divided into quarters. Each of two sides of the
square represents a pair of gametes. The probabilities
(that is, frequencies) of each possible combination of
gametes can be worked out from the divisions of the square.

Thus, the results of crossing the parents to produce the F_1
can be shown as in *figure 5*.

Figure 5
This shows the results of crossing
parent flies to produce the F_1
generation. Notice that it does not
matter if the position of the male
and female gametes is reversed,
nor if the sex of the parent flies is
reversed.

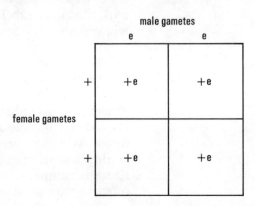

The perpetuation of life

All the F_1 progeny will be the same, that is, $+e$.

The results of crossing members of the F_1 to produce the F_2 can be shown as in *figure 6*.

The probabilities of each type of offspring in the F_2 will be:

$$+ + = \tfrac{1}{4}, +e = \tfrac{2}{4} = \tfrac{1}{2}, ee = \tfrac{1}{4}$$

Figure 6
Here the results to be expected from crossing members of the F_1 generation are seen. Again notice that the position of the male and female gametes can be reversed without altering the results.

Whatever method of presentation is used, the diagrams should not become too overbearing and their symbols should be clearly shown. Diversions, such as the discussion of other characteristics, should be avoided.

It is suggested that, at this stage, it is unnecessary to use the term genotype. The distinction between genotype and phenotype is considered in Chapter 9. Students may be concerned about the distinction between organism and zygote. This is referred to in section 3.2.

2.4 Using models

Objectives

1 to devise a model of inheritance by using beads, to help to explain the results of breeding investigations

2 to introduce the concept of probability

Students may not be used to the term 'model' as defined in this section. To them a model is likely to be either a replica of an object (for example, a model railway or a dressmaker's model) or a reference to some ideal condition, as in the phrase, 'he is a model of good manners'. (It may also mean a fashion model – the glamorous women on the front of magazines.)

Using models is an important aspect of scientific work. It is referred to on several occasions in the course and it is essential that students understand the way scientists use the term as early as possible.

Students will probably work in pairs or larger groups for this exercise.

Apparatus and materials
Per pair or group:
200 beads of one colour (*e.g.* red)
200 beads of another colour (*e.g.* yellow)
5 containers for beads (*e.g.* plastic beakers or jam jars)

Breeding with beads

In the investigations, single beads are used to represent separate gametes and pairs to represent zygotes. After the work of Chapter 3, students will be able to appreciate that single beads represent genes and pairs genotypes. Ideally, large numbers of beads should be used to represent gametes and the chosen pairs of beads (zygotes) should be scored, returned to their respective 'gamete containers', and shaken. In this way the probability of selecting each type of zygote is not influenced by the gamete previously removed.

Q1 As a result of this model investigation, predict the type (or types) of flies you would expect to find in the F_1 generation of *your* breeding investigation.

The beads representing the zygotes (= organisms) of the F_1 would be in similar pairs. Therefore you would expect the flies in the F_1 generation to be all the same.

Q2 Suppose the grey-bodied flies are female and the ebony-bodied flies male. Would this make any difference to the results you would expect?

No. The beads representing the F_1 come out just the same.

Q3 What type (or types) of flies would you expect to find in the F_2 generation of *your* breeding investigation?

For every four flies, probably one would be like one of the parents, one would be like the other parent, and the other two would be like the F_1.

Q4 How can you use these model investigations to help to explain your results?

The apparent disappearance of the ebony body characteristic in the F_1 and its re-appearance in the F_2 can be explained by assuming that the factor for grey body is dominant over the ebony body factor when both occur together and that in the zygote there is always a pair of factors. With these assumptions the bead model shows how the inheritance of these characteristics may take place.

The next question to ask is 'What form do these "factors" take in the organisms and their gametes?' This leads on to the work of Chapter 3.

Class exercises with beads can be done very quickly by distributing the total number of beads for an exercise among members of the class. Larger numbers than those stated may be used, but they should not be smaller. The results obtained by individual students can be compared with class results and will inevitably show the influence of sample size on the relation between expected and observed results. This is a similar exercise to 'Some problems dealing with probability' in this section in the *Text*.

Coloured beads are particularly useful for work in population genetics. In Chapter 9 full information can be found.

Probability

At this stage of the course two points need to be made understandable to students.

1 The probability of an event taking place is the fraction of the number of ways the particular event can occur divided by the number of all possible events. In other words, if an event, A, can occur in m cases out of a total of N equally likely cases, then the probability, P, of the event taking place is:

$$P = \frac{m}{N}$$

The production of ebony-bodied flies in the F_2 of the breeding investigation can thus occur in one case out of a total of four equally likely cases of fertilization. If A is the production of ebony-bodied flies,

$$m = 1, N = 4$$
and therefore $P = \frac{1}{4}$

The probability of producing ebony-bodied flies is $\frac{1}{4}$, so that it is to be *expected* that one out of every four flies will have an ebony body.

2 The larger the number of cases that are taken – the larger the number of F_2 flies raised – the easier it is to see if the expected fractions are being realized.

For later work, especially with students who have a mathematical bent, this account of probability could be broadened to include cases in which probabilities are combined. The following rules could be dealt with:

3 The probability that some one of a set of mutually exclusive events will occur is the *sum* of the probabilities of the single events. For example, the production of flies of normal body colour (grey) can occur in three cases out of a total of four equally likely cases of fertilization. The probability of each case is $\frac{1}{4}$. Therefore the probability that one act of fertilization will produce a fly of normal body colour is $\frac{1}{4} + \frac{1}{4} + \frac{1}{4} = \frac{3}{4}$. That is, it is to be *expected* that three out of every four flies will have normal body colour.

4 The probability that all of two or more independent events will occur is the product of the individual probabilities. Two or more events are said to be independent if the

occurrence of one of them does not affect the occurrence of the others.

a The probability of producing a homozygous fly of normal body colour in the F_2 is $\frac{1}{4}$. The probability of producing a homozygous ebony-bodied fly is $\frac{1}{4}$.

The probability that of the first two flies which hatch, the first will be a homozygous fly of normal body colour and the second a homozygous, ebony-bodied fly will be $\frac{1}{4} \times \frac{1}{4} = \frac{1}{16}$. That is, it is to be expected that one pair out of 16 successive pairs of flies hatched will consist of, first a homozygous fly of normal body colour and then, a homozygous, ebony-bodied fly.

b Similarly, the probability that of the first two flies which hatch, the first will be a homozygous, ebony-bodied fly and the second a homozygous fly of normal body colour will be $\frac{1}{4} \times \frac{1}{4} = \frac{1}{16}$.

From the definition of probability, if an event is certain to happen its probability is 1. If it is certain not to happen its probability is 0. Probabilities can therefore be expressed as vulgar fractions or as decimals between 0 and 1. Thus, a probability of $\frac{1}{4}$ can also be expressed as 0.25.

Some problems dealing with probability

Q5 What is the probability of drawing an ace from a pack of playing cards?

Probability of drawing an ace $= \frac{4}{52} = \frac{1}{13}$.

Q6 Ten red beads, five yellow beads, and two green beads are in a container. What is the probability of drawing out a yellow bead? a green bead?

Probability of drawing a yellow bead $= \frac{5}{17}$.
Probability of drawing a green bead $= \frac{2}{17}$.

Q7 The total bird population of an area consists of 5000 individuals, of which 2500 are sparrows and 1000 are chaffinches. What is the probability that if you see one bird it will be a sparrow? What is the probability that it will be neither a sparrow nor a chaffinch?

Probability of seeing a sparrow $= \frac{2500}{5000} = \frac{1}{2}$.
Probability of seeing neither a sparrow nor a chaffinch $= \frac{1500}{5000} = \frac{3}{10}$.

Q8 The probability of a man being colour-blind is 1 in 25. In a crowd of 1600 men, how many are likely not to be colour-blind?

Probability of a man not being colour-blind $= \frac{96}{100}$. Therefore, the number of men out of 1600 likely not to be colour-blind $= 1600 \times \frac{96}{100} = 1536$.

Q9 Place 100 red and 100 yellow beads in a container. What is the probability that a red bead will be taken from the container?
a Draw out ten beads, one at a time. How many are red and how

The probability of a red bead being taken from the container $= \frac{1}{2}$. If ten beads are drawn out, five would be expected to be red and five yellow. The results of the investigation will show how larger samples reproduce the proportions expected more accurately.

The perpetuation of life

many are yellow? Replace them
and shake the container.
b Draw out an additional 10 beads.
What is the *total* number of red
beads taken out of the container?
How many are yellow?
c Repeat this procedure a number
of times. Each time take out 10
beads and note the *total* number of
red and yellow you have from
carrying out *a*, *b*, and *c*, as well as
the *individual totals* of red and
yellow for each batch of ten.
What conclusion would you draw
from this investigation?

2.5 Investigations with other organisms

Objective

To establish the general truth of
the pattern of inheritance
investigated in *Drosophila*

Although *Drosophila* is probably the best organism to use
in this work, students should realize that similar results
can be obtained with other organisms. They should also
appreciate that breeding takes much longer with most
organisms than with *Drosophila*.

At least one long-term investigation should be set up using
an organism other than *Drosophila*. Suggestions about
long-term work are given in the Preface.

Two types of crosses could be dealt with at this stage:
1 To show monohybrid inheritance in which dominance is
displayed.
2 To show monohybrid inheritance in which there is a
distinct heterozygote due to the absence of dominance.

Symbols
When students reach the stage of analysing the results
of long-term investigations, they will be familiar with the
word 'gene' and should also understand the conventions
by which symbols are used to designate different genes.

The symbol for a 'wild type' or normal gene is +. This can
refer to all wild type genes or to a particular gene,
depending on the context. If an organism is homozygous
for a wild type gene, it is sufficient to put +. However, for
students working out breeding problems it is often helpful
to use + +.

Mutant genes are symbolized by letters. Dominant mutant
genes are designated by capitalizing the first letter of
the symbol, as in Sd or A. Recessive mutants are designated
by small letters, as in se, p, and vt or we.

When organisms are designated only by mutant symbols it is assumed they are homozygous for the characteristics referred to and contain no other visible mutants. Again, however, students working out breeding problems may find it helpful to use double symbols, e.g. sese or AA.

If an organism is heterozygous, the symbols for both genes should be given as in +/se, the phenotype being wild type. If several genes, some of which are mutants and others wild type, are being designated, specific wild type alleles can be indicated by a small + above and to the right of the mutant symbol. For example, se^+ is the wild type allele for se. There are further conventions for denoting the position of genes on different chromosomes. Information on this can be obtained from books listed under 'Reference material' on page 42.

Mice

(For the care of mice see Appendix 2, page 213.)

The following crosses will each give an F_2 showing a monohybrid ratio (3:1) in which dominance is displayed.

The dominant gene is placed on the left. 'Normal' is signified by +.

Coat colour
    agouti (A) × non-agouti (a)
    black (B) × brown (b)
    intense (+) × dilution (d)

Coat colour can be classified eight days after birth. For easy classification of the last two crosses, it is preferable that the organisms should be homozygous for non-agouti.

Eye and coat colour:
black-eyed intense (+) × *pink-eyed dilution* (p)
Eye colour can be classified from birth.

Ear size: normal (+) × *short* (se)
Ear size can be classified 18 days from birth.

Hair: wavy (we *or* wa) × *normal* (+)
Hair texture can be classified 4 days after birth (whiskers first, coat later).

Monohybrid inheritance without dominance (1:2:1) can be investigated by breeding up to the F_2 with the following cross: coat colour: agouti (A) × tan belly (at) or extreme chinchilla (c^e) × albino (c).

Mice for genetical work can be obtained from biological supply agencies.

Plants

Adult characteristics

Investigating the adult characteristics of most plants is slowed down because the length of the breeding cycle limits breeding to a generation each year. With a few plants, however, two or sometimes three generations can be obtained in a year. By using suitable light and heat, an investigation can be made to cover the greater part of the year. This work is best started indoors in late January or February. For details of methods of rearing and hand-pollination, see Appendix 2, page 206.

1 *To demonstrate monohybrid inheritance with dominance,* the following species are suitable:
Pisum sativum (edible pea); tallness dominant to dwarfness.
Lathyrus odoratus (sweet pea); tallness dominant to dwarfness.
Seeds of both edible and sweet peas (tall and dwarf varieties) can be obtained from most biological supply agencies and nurserymen.

These investigations with peas are basically a repetition of Mendel's work. The heterozygotes can be selfed by covering the flowers with a bag. Care must be taken in handling the flower buds as they easily fall off. It is usually only possible to obtain one generation a year.

Antirrhinum majus (snapdragon): rust resistance dominant to rust susceptibility. The varieties, Sutton's Victory (rust resistant) and Sutton's Eclipse (rust susceptible), are suitable and obtainable from nurserymen and biological supply agencies. One generation per year can be obtained. Indoors, antirrhinums can be kept as perennials; outside, they are best treated as annuals.
Rust can usually be obtained from specimens of Sutton's Eclipse grown out of doors. To test for rust resistance, place infected leaves in a jar of water and stir well to distribute the spores. Spray the water onto plants with a fine syringe (a cool period in the late afternoon or early evening is the best time). If rust fungus does not develop in a week or two, the plants can be considered to be resistant.
This investigation can be used as a basis for the discussion of the inheritance of susceptibility to disease.

Q1 Can you predict the presence of factors D and d in the resistant and susceptible rows of barley (*figure 26*)?

The rows that are susceptible to mildew must be dd (d is recessive). The rows that are resistant to mildew could be DD or Dd.

2 *For demonstrating monohybrid inheritance without dominance* it is convenient to use:
Senecio vulgaris (groundsel): single gene inheritance for tubular and ray florets. The heterozygote has intermediate florets.
The common groundsel has tubular florets, its success being due mainly to its greater reproductive capacity. To self the heterozygotes, cover the flowers with a bag.
Two, or possibly three, generations may be obtained in a year.

Human inheritance

Suggested investigation
1 Construct your own family tree as far back as you can.
2 Find out the colour of the eyes of your relations.
This may entail some discussion and, if you have time, letter writing. You may have to rely a lot on people's memories.
3 Show the distribution of the different eye colours in your family tree, using these symbols:

● blue-eyed female ■ blue-eyed male
○ non-blue-eyed female □ non-blue-eyed male

By non-blue we mean any colour of eyes except blue.

Here are some suggested questions:
'What possible errors may there be in your results?'

(Those arising from the smallness of the sample studied and unreliable judgments of the colour of the eyes of people with whom personal contact is not possible.)

'Does the inheritance of eye colour follow the same pattern as the inheritance of body colour in *Drosophila*?'

(With the limitations mentioned above, yes. Instances can be found of the blue eye characteristic apparently disappearing in a generation.)

'Can the inheritance of eye colour be explained in the same way as with body colour in *Drosophila*?'

(Yes. There must be two 'factors' in the zygote (or organism) and one in each gamete.)

Personal difficulties
In dealing with human inheritance teachers must be prepared for the exceptional but inevitable occasions on which the discussion becomes a source of embarrassment to a student. Sometimes there may be reference to a characteristic which marks out one student in the class. Sometimes a student may have to acknowledge his

illegitimacy – other students may not know. At the same time, it should be borne in mind that apparent difficulties may be caused by a misunderstanding of genetic mechanisms.

Some of these difficulties can be anticipated by a knowledge of the students' backgrounds and it is always wise to glance at individual results of investigations before they are brought to the attention of the whole class.

The possibility of over-simplification
It is proper to stress this in all work dealing with human inheritance, and this should be done *before* the work is performed, and in time to minimize possible social embarrassment for a student. Mutation can be used to account for genetic abnormalities only rarely, and at this stage in the course is best left as an unexplained complication.

Human eye colour. The inheritance of eye colour in man is more complex than was once thought. In general we can consider that blue or grey eyes are the result of a double recessive condition. For our purpose, darker, non-blue eyes, such as brown, hazel, and green can be considered to be the result of the influence of a dominant gene, though a whole series of genes probably modifies the action of a primary eye pigment gene.

2.6 Inheritance in seedlings

Objective

To demonstrate further the pattern of inheritance investigated in *Drosophila*

Apparatus and materials
packaged genetic lesson (PGL 15 or PGL 5 is recommended)
7 seed trays (approximately 35 cm × 22 cm). *1* tray is required for each packet of seeds except F_2 and backcross which need *2* each
seed compost for trays
plant labels

The work with *Drosophila* can be rather long and fragmented. For some students this may be an advantage but for others it will mean that they may lose the thread linking ideas together. For this reason, and because it is simple and effective, a practical investigation with quick impact is recommended.

Various genetic types of seed can be obtained which show clear differences in their seedlings. Biological supply agencies are one source. Another is Practical Plant Genetics, 18 Harsfold Road, Rustington, Sussex. This produces 'Packaged Genetic Lessons' (PGL) with seeds of parents, F_1, F_2, and backcross normally supplied together with full notes on sowing, maintenance, and analysis of results. Most of the PGL use tomato seedlings but PGL 15 concerns cotyledon flavour in cucumbers. One type of parent has bitter-flavoured cotyledons (dominant). PGL 15 includes a packet of seeds of each parent type and F_2. Students can sow these seeds and then test for cotyledon flavour in a 'tasting session' five to seven days later. Careful planning of the tasting session and the collection of class data will be essential.

Alternative lessons are PGL 2 (stem–hypocotyl–colour) and PGL 10 (stem hairiness). PGL 4 (cotyledon colour) shows incomplete dominance giving $1:2:1$ in the F_2. However, you may prefer to keep PGL 4 for the work in Chapter 5 on the influence of environment on development (5.2). PGL 5 (stem colour and leaf shape) is usually used to show dihybrid inheritance but it could be used here as a 'double test'. Stem colour can be scored after about one week, while leaf shape only becomes clear about a week later. Students will usually score $3:1$ for both separate 'tests' and only a few are likely to develop the idea *independently* of considering the inheritance of stem colour *in relation* to the inheritance of leaf shape. All the packaged tomato lessons include seeds of both parents, F_1, F_2, and backcross. Quantities are sufficient for a class of 30 or for a series of parallel classes if the seedlings are protected from unnecessary damage. This is also true of PGL 15, where the cotyledons are large enough to be 'sampled' a number of times for bitter taste.

Students could sow the seeds or, if time is severely limited, seedlings could be presented as a test. Again, careful planning of scoring and collecting data by the class is essential.

Where seedlings are not available colour plates 6, 7, and 8 in the *Text* will help students to test their understanding of the principles of inheritance.

Q1 What character differences can you identify?

The answer depends on the material provided. The colour plates, 6, 7, and 8, show:
Maize cobs: coloured seeds and white seeds
Tomato seedlings: purple stem (hypocotyl) and green stem
Tomato seedlings: cotyledon colour; dark green, pale green, and yellow.

Q2 Devise a scheme to explain the ratio of different types seen, indicating the factors carried in the other generations.

The answer depends on the material provided.
Note: In maize seed colour there are two pairs of genes involved but here we could expect the following:
C = colour   c = white
Tomato hypocotyl colour: A = purple stem
                          a = green stem

Tomato cotyledon colour: G = green   g = yellow
G and g are co-dominant so Gg plants are pale green. An
exercise involving these tomato seedlings is suggested for
Chapter 5 in section 5.21. A scheme is given in answer to
question 4 of section 5.21, on page 85 of this *Guide*.

Suggestions for homework or preparation

Some of the problems in the *Text* which do not require
apparatus, etc., could be set as homework.

Other problems

1a 'A Japanese girl with dark brown eyes and black hair, of
pure Japanese ancestry, married a fair-haired American of
Swedish ancestry who has blue eyes. All their 10 children
have dark hair and brown eyes. These children in turn
marry and have 32 children between them. Most of these
children have brown eyes but a few have blue eyes.
What can be inferred from these data about the nature of
eye-colour inheritance?'

Brown eye is dominant to blue eye. There are two factors in
the zygote: one in each gamete.

b 'One of the F_1 children of this family marries a man with
dark hair and they have 6 children. All the children have
dark hair but 3 have blue eyes and the other 3 have
brown eyes.
What can be inferred from these data about the colour of
their father's eyes?'

The inheritance of eye colour is not related to hair colour.
The 50:50 ratio of blue-eyed to brown-eyed children suggests
that the father was homozygous for blue eyes. Thus if
B = brown eye (dominant), b = blue eye (recessive):

Mother (F_1 – Bb)

		B	b
	b	Bb	bb
Father (bb)			
	b	Bb	bb

Therefore, probability of children having blue eyes
$= \frac{2}{4} = \frac{1}{2}$.
Probability of children having brown eyes $= \frac{2}{4} = \frac{1}{2}$.

Therefore, the observed result is the same as the expected
result.

However, the small number of children involved does not permit this to be established as a firm conclusion.

2 'The proportion of the sexes of the children born in human populations tends to be about 50:50. Assuming maleness and femaleness are the result of single-factor inheritance, what are the genotypes of a man and woman likely to be?'

To obtain a 50:50 ratio in the offspring, one parent must be homozygous and the other heterozygous. Using the usual symbols, sex will thus be inherited:

		Female	
		X	X
Male	X	XX	XX
	Y	XY	XY

Probability of a child being male $= \frac{2}{4} = \frac{1}{2}$.
Probability of a child being female $= \frac{2}{4} = \frac{1}{2}$.

3 'Suppose you were given a male and a virgin female *Drosophila* in a culture bottle. You know them to be either both + + or +e.

a What investigation could you carry out to determine that they were using no other stocks of fly?

Allow them to mate and remove them when larvae are present.

b 'How would the results of the experiment determine the answer?'

All offspring of normal body colour indicate + + parents. If a quarter of the offspring are ebony, this means +e parents.

4 'Suppose you are an astute farmer. You are offered a herd of black cattle for a small sum because, for some reason, black cattle are inferior to brown. By chance you hear that the parents of the herd were a black bull and brown cows, all of whom were pure bred. Would you buy? Why? If you did buy, what breeding scheme would you introduce?'

It appears that coat colour acts as a 'marker' and is 'linked' with desirable or undesirable qualities in the cattle. Assuming this, the problem can be treated as a case of single-factor inheritance.
Thus, if B = dominant black, b = recessive brown. The parents of the herd would be BB (bull) × bb (cows). The herd would be Bb.

If there were a bull in the herd, it could be crossed with the cows in it, that is, Bb × Bb. This would produce 1 in 4 brown cattle (bb) which could form the basis of a breeding scheme. If there were no bull the cows could be sired by a pure bred brown bull, i.e. Bb × bb. This would produce 1 in 2 blacks (Bb) and 1 in 2 browns (bb). These again could be the basis of a breeding scheme. A bull is likely to be included among the offspring.

Background reading

Q1 Which of the investigators designed and used controls in their work?

Q2 Is the way life is believed to have started an example of spontaneous generation?

Q3 How can scientists today investigate this problem of the origin of life?

Spontaneous generation

Redi, Pasteur, and Spallanzani.

For discussion. Emphasize the point that the origin of life may have been spontaneous generation occurring once only.

By simulating, in a special apparatus, the conditions believed to have been present on Earth many millions of years ago. It might be a good idea to illustrate discussion with a diagram of Miller's pyrosynthometer (*figure 7*).

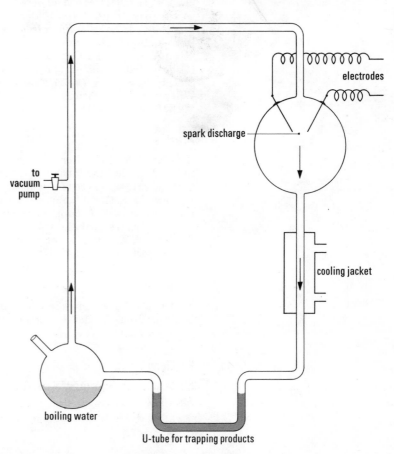

Figure 7
Miller's pyrosynthometer. The diagram shows the apparatus used by Stanley Miller in 1953 to synthesize organic compounds by simulating primitive Earth conditions believed to exist about 3×10^9 years ago. After evacuation the apparatus was filled with hydrogen, methane, and ammonia. As a result of discharge, products were collected in the U-tube and Miller identified some 15 different amino acids: the 'building blocks' of protein.
After Miller, S. L. (1955) 'Production of some organic compounds under possible primitive Earth conditions', Journal of the American Chemical Society, **77**.

electrodes

spark discharge

to vacuum pump

cooling jacket

boiling water

U-tube for trapping products

Summary

At this point, students may be asked to make their own summary of the work of this chapter. Alternatively if you have been doing Chapter 3 before completing this chapter, students should try to relate the results of breeding investigations to chromosome behaviour at meiosis.

Reference material

Books

*Reading suitable for students

Auerbach, C. (1965) *Heredity, an introduction for O-level students.* Oliver & Boyd. (A very good text. Most suitable as background reading.)
Bailey, N. T. J. (1968) *Statistical methods in biology.* English Universities Press. (Written for those biologists who have little mathematical skill.)
Bernstein, L. L., and Weatherall, M. (1951) *Statistics for medical and other biological students.* Livingstone. (A good chapter on probability.)
Crane, M. B., and Lawrence, W. J. C. (1952) *The genetics of garden plants.* 4th edition. Macmillan. (Useful for obtaining ideas for long-term investigations.)
Darlington, C. D., and Bradshaw, A. D. (Eds) (1963) *Teaching genetics.* Oliver & Boyd. (Useful articles are: Crowe, L. K., 'Seedling characters' page 90; Falconer, D. S., 'The use of mice in teaching genetics' page 44.)
*Haskell, G. (1961) *Practical heredity with* Drosophila. Oliver & Boyd. (An introductory account.)
Moroney, M. J. (1969) *Facts from figures.* Penguin. (A useful introduction to statistics.)
Penrose, L. S. (1963) *Outline of human genetics.* 2nd edition. Heinemann. (Concise and readable: strongly recommended.)
Strickberger, M. W. (1962) *Experiments in genetics with* Drosophila. Wiley. (Contains virtually all that one wants at all levels.)
Wallace, M. E. (1971) *Learning genetics with mice.* Heinemann.

Articles

Brierley, J. K. (1961) 'Some suggestions for the teaching of evolution in the field, garden and laboratory'. *School science review*, **42**, *148*, 401–10. (This describes the use of a genetic garden.)
Wallace, M. E. (1963) 'Laboratory animals: cage design principles, practice and cost'. *Laboratory practice*, 354–9. Animal Technicians Association. (Describes the use of the 'Cambridge' cage in detail.)
Wallace, M. E. (1965) 'Using mice for teaching genetics'. *School Science review*. Part I, **46**, *160*, 646–58; part II, **47**, 39–52.

16 mm film

'Genetics and plant breeding', sound, colour, 17 minutes. 16 and 35 mm. Unilever Films, Unilever Film Library, Unilever House, London EC4P 4BQ.

Film loop

Nuffield O-level Biology 'Handling *Drosophila*'. NBP–70. Longman.

Objectives of this chapter

1 to introduce the logical hypothesis that gametes must be concerned in inheritance

2 to devise a model of inheritance assuming that the nucleus is responsible for inheritance

3 to study cell division in growing tissues and examine the essential features of mitosis

4 to study gamete formation

5 to introduce the essential features of meiosis

6 to compare chromosome behaviour with the way factors 'behave' in the breeding investigations

7 to arrive at the logical conclusion that the chromosomes are the physical site of inheritance

3.1 A living thing begins

Objective

To introduce the logical hypothesis that gametes must be concerned in inheritance

Apparatus and materials
Per pair or group:
living specimens of *Pomatoceros* (attached to stones) in sea water
microscope
3 cavity blocks (or watch-glasses)
2 dropping pipettes
microscope slides
coverslips
50 cm^3 of sea water
scalpel (or awl used in mammal dissection)
blunt seeker

This practical investigation usually works well all the year round but is more certain in the early months of the year. Teachers may prefer to do this work as a demonstration. The main problem is likely to be speed. Students should be encouraged to work fairly quickly to remove worms from their tubes. If the practical investigation is not possible or not a success, show the film loop 'Fertilization in the marine worm *Pomatoceros triqueter*' (see the Reference material on page 58).

Q1 Can you see the nucleus of the egg? Make a sketch of those features you can determine.

With suitable lighting the nucleus will be visible in some eggs. Details of nuclear structure will not be seen. This is important because some students may well be familiar with the idea of chromosomes in a nucleus and expect to see them.

Q2 How large are the sperms in relation to the egg? Make a sketch.

The important point is that students should recognize the great difference in the size of gametes.

Q3 How many sperms can you find around one egg? What are they doing?

The large number of sperms and their activity should be clear.

3.2   A model of inheritance

<table>
<tr><td>Objective</td></tr>
</table>

To devise a model of inheritance assuming that the nucleus is responsible for inheritance

This section of the chapter is an important chance to set up a clear hypothesis and then test it in the following practical work. Teachers will need to refer to this model again in section 3.5 when summarizing the work of Chapters 2 and 3.

From the model we lead logically into the two questions:

a   How does the fertilized egg ($2n$) pass on two nuclear factors to each of its many daughter cells as it grows, so that the cells of the adult organism will be $2n$?

b   How do the cells in the reproductive organs of the adult organism ($2n$) divide to produce gametes which carry only one nuclear factor (n)?

3.3 Cell division in growing tissues

<table>
<tr><td>Objective</td></tr>
</table>

To study cell division in growing tissues and examine the essential features of mitosis

Apparatus and materials
(required for practical work with root tips)
Per pair:
young onion root tips fixed in acetic alcohol ($\frac{1}{2}$ hour minimum but no upper time limit) and stained in Feulgen ($\frac{1}{2}$–2 hours)
single-edged razor blade
pair of forceps
4 microscope slides
coverslips
2 dissecting needles
filter paper
Bunsen burner or spirit lamp
microscope

Students are provided with stained root tips of onion ($2n = 16$). See 'Roots' in Appendix 2, page 219, for details of culture. These can be grown at any time before the practical work and fixed in acetic alcohol (1 part glacial acetic acid; 3 parts ethanol). The root tips are then hydrolysed in 1M hydrochloric acid at 60 °C (water bath) for six to eight minutes and placed in Feulgen stain, at room temperature, in the dark. They should be left in the stain for a period of thirty minutes to two hours. This hydrolysis and staining must be carried out immediately before the lesson. Successfully stained root tips will be magenta. Should this staining technique fail on the day, because of some technical error, use methylene blue on the same or other roots (see page 194). Squashing the root tips should present no special difficulty and warming the slide over a low flame is not essential.

Q1 Estimate what proportion of an average cell found at the root tip is occupied by the nucleus.

Estimates will vary widely but the nucleus is seen to be relatively large.

Q2 Is there any difference in this respect between cells near the tip and those farther away?

Yes, the nucleus remains about the same size but the cells are larger away from the tip.

Q3 In which cells are chromosomes visible?

Chromosomes are most likely to be seen in cells near the tip where division is most frequent.

Q4 How are the chromosomes arranged in the cells?

Q5 Construct drawings of the different types of cells you can detect. Compare them with the photographs of similar cells (*figure 36*).

Students will probably see a variety of stages and may make sketches of metaphase and anaphase.
There is no need to use the terms metaphase, etc., and teachers should not reveal the sequence of mitosis. This is the subject of the next questions based on the photographs in figure 37 in the *Text*.

Q6 The photographs of cells from the root tip of *Aloe* have been labelled *a* to *i* and their sequence is random except that *a* shows a nucleus just before cell division and *i* shows the two daughter nuclei after cell division. Using the letters of the photographs, write down the probable sequence of events during cell division.

$a, c, g, b, d, h, f, e, i.$

Q7 How many chromosomes are there in an Aloe root cell? Let this number equal 2n.

$14 = 2n.$

Q8 Can you divide the chromosomes by size into two groups?

Yes.

Q9 How many chromosomes are there in each group?

Eight large or long ones, six small or short ones.

Q10 Is each chromosome in *figure 40* seen as a single or double thread?

Double.

Q11 Is each chromosome in *figure 37h* seen as a single or double thread?

Single.

Q12 How many chromosomes are about to enter each of the two daughter cells?

14.

Q13 How does this number relate to the value for 2*n* given in answer to question 7?

It is the same. A 2*n* parent cell has given rise to two daughter cells *each having 2n chromosomes*.

Q14 How many long chromosomes and how many short are going into the 'top' cell and how many into the 'bottom' cell?

Eight long and six short are going into each daughter cell.

Q15 Summarize the events of normal cell division (called *mitosis*) to explain the number of chromosomes in each parent cell (*figure 40*) and the number of chromosomes entering each daughter cell (*figure 37* and *figure 41*). Use the idea that the normal number of chromosomes is equal to 2*n*.

1 Chromosomes appear as two threads (chromatids).
2 One thread of each chromosome goes into each daughter cell.
3 Each daughter cell inherits an exact replica of chromosome material from the parent cell.
4 Each daughter cell has the same number of chromosomes (2*n*) as the parent cell.

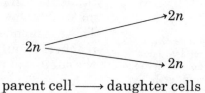

parent cell ⟶ daughter cells

Phase	Time (minutes)
prophase	71
metaphase	6.5
anaphase	2.4
telophase	3.8

Table 3
Duration of mitotic phases in onion.

Students often ask about the length of time required for mitotic cell division. *Table 3* gives an estimate of the time taken for the phases of mitosis in onion. This makes it clear that even in actively dividing tissues most cells will be seen in prophase and very few in anaphase.

Teachers might like to try growing *Aloe* and use actively growing roots for mitosis preparation. Students could then compare their results directly with the photographs.

The following notes describing the photographs of mitosis have been prepared by Dr P. E. Brandham.

'The mitosis series'
'Figure 37 in the *Text* uses Feulgen squashes of root tips.
a Interphase. Small meristematic cells each containing one relatively large nucleus. Nucleoli visible as non-staining "bubbles" inside the nuclei.
c Prophase (early). Chromosomes becoming visible as a mass of intertwined threads.
g Prophase (late). Chromosomes contracting. Just recognizable are eight long ones and some short, one centro-

mere clear. Longitudinal division into chromatids in several places.

b Metaphase. Chromosomes fully condensed. Centromeres gathered on equator of non-stained spindle. Arms of chromosomes lie above and below it. Eight long chromosomes clearly seen.

d Early anaphase. Chromatids of each chromosome separate and move to poles, centromere first. The centromeres are all near one end in this plant, so the chromatids at this stage are always hooked. Short chromosomes are stiffer and do not hook.

h Mid anaphase. All chromatids nearing the poles. Eight long chromosomes clearly visible on each side.

f Late anaphase. Chromatids reach the pole and begin to clump together.

e Telophase. Chromatids have lost their individual identity, and have contracted into two small dense nuclei. The new cell wall starts to form on the equator of the spindle.

i Late telophase–interphase. The nuclei gradually expand to resemble typical interphase nuclei. During this stage the DNA is duplicated in preparation for the next mitosis.

'**Figure 38** in the *Text* shows a root tip treated with alpha bromonaphthalene before fixation and staining. This contracts the chromosomes more than during mitosis and also prevents spindle formation so that the chromosomes lie loosely in the cell. They will then spread out when the cell is squashed. Each chromosome has a centromere near one end (in the acrocentric position). Two pairs of chromosomes have another non-staining region, the secondary constriction which is responsible for nucleolar organization. Beyond this constriction is a small satellite. Eight long chromosomes and six short can be seen. (This is a diploid number.)'

3.4 Cell division in making gametes

Objectives

1 to obtain firsthand information of the problems in studying meiosis, by a practical investigation of locust testes

2 to determine the essential features of cell division in making gametes (meiosis), by investigating a series of photographs showing meiosis in *Aloe*

Examining locust testes

Through actually dissecting the locust testes and making a squash preparation, students should get some idea of the size of the cell and the chromosomes. By proceeding down in size from locust to testis and then to cell, it is hoped that they will become aware of the smallness of the chromosomes and get some idea of their spatial relations with a whole organism.

The practical work with the locust is a good exercise in technique but results are not always good. It should not be seen as an essential part of the main theme and the photographs of meiosis in *Aloe* should enable students to make an *investigation* from secondhand data.

Demonstration of locust dissection and squash technique
Apparatus and materials
Per pair:
locust (male, fifth instar or young adult)
4 glass slides and coverslips
Bunsen burner or spirit lamp
orcein stain and pipette
saline solution and pipette
filter paper
scalpel, or brass rod with blunt end, or similar instrument
pair of forceps
2 mounted needles
12 pins
cork mat or waxed watch-glass
hand lens
microscope

It is suggested that this demonstration should be done in three phases:

1 The teacher dissects and prepares a slide of locust testes, explaining the stages as he proceeds.
2 Show the film loop 'Squash preparation' (see the Reference material on page 58).
3 The students perform the technique in stages, and each stage is linked to the instructions in the *Text*. This is done under the direct guidance of the teacher. Students work in pairs.

For the demonstration, preserved or fresh adult locusts can be used, as the aim is only to instruct the students in the techniques as such.

The most difficult problem for the students is the identification of the testes of the locust. A permanent preparation showing the testis partly disentangled from the fatty material is a useful asset. It is probably best to show this to individuals who find identification difficult. Teachers may prefer to provide students with small pieces of locust testis.

A few good permanent preparations showing locust meiosis should also be available for assisting students who have difficulties in interpreting their own preparations.

A pointer eyepiece is a valuable aid for demonstrating material under a microscope to individual students. It consists of a $\times 6$ or $\times 10$ eyepiece in which is fitted a small movable pointer operated by the finger. A particular place on a slide can then be easily indicated to a student.

It may also be necessary to remind the class about the use of the microscope, particularly high-power. A film loop on the microscope (such as 'The use of the microscope', listed on page 58), together with the instructions given in the *Text* would be of assistance.

A few extra locusts should be available for replacement.

The work is best performed in pairs. The dissection and microscopic examination are probably best done as joint exercises by each pair. Once the testes have been extracted they can be divided in two and each student can make one or more squash preparations.

While the students are examining and drawing their preparations, brief discussions with groups of about six students can be held. Points worth considering are:
1   That the basic technique they are using is similar to that used by research workers.
2   The relative sizes of the body of the locust, the testis, the cell, and the nuclei.
3   The interpretation of the arrangement of the parts of the nuclei (= chromosome).
4   The importance of drawing, boldly and clearly, only what is relevant.

The preparations the students make may show them that at times the nucleus is divided into rod-like or cord-like elongated parts. Some of them may show that they are found both in the centre of the cell and in two groups towards opposite ends of a cell. This is as much as one would want the students to achieve with this work.

It is not to be expected that every student will make a satisfactory preparation. Possibly 50 per cent of the students may do so. The important thing is to ensure that each student knows how one is made.

Locusts
Both the desert locust *Schistocerca gregaria* and the migratory locust *Locusta migratoria migratorioides* are suitable for this work. Unless locusts are kept permanently at your school it is best to arrange for the ones you order to arrive a few days before they are to be used. The date on which they are to be used should be clearly mentioned when ordering. (For arrangements for temporary culture, see Appendix 2, page 210.)

Some meiotic divisions may be found in locusts of quite a wide range of ages – from fifth-instar nymphs to fairly old, yellow adults. It would appear that the best age to use to obtain large numbers of cells undergoing meiosis is the young adult, within one or two weeks of its last moult.

Locusts are used in this work because they are easily obtained and provide material that allows students to be reasonably successful with their preparations. Another reason is that students should find their results more credible if, the first time they compare meiosis and the pattern of inheritance shown by breeding, they can do so through work performed with similar organisms.
Locust and *Drosophila*, both being insects, appear to be the best combination obtainable.

Only testis material is suitable for meiotic chromosome squash preparations, so only males will be required.
Male. The posterior tip of the abdomen has a distinctive rounded appearance rather like the stern of a boat. This is due to an upward projection of the lower plate of the last abdominal segment.
Female. The female is usually rather larger than the male. Her abdomen is terminated by somewhat jagged tooth-like structures which form an ovipositor.

Figure 8
Lateral views of the abdomens of the male and female locust. The male is above.
Photograph, John Myers.

The perpetuation of life

Killing locusts. The locusts are best killed shortly before the class is taken. Any vessel in which filter paper or cotton wool soaked in chloroform–ether can be suspended, will do as a killing chamber. (The soaked material should *not* be dripping.) The locusts are best kept in the vessel for ten minutes. If many locusts are killed together they often get covered with faeces. These should be washed off before they are dissected.

Q1 What form does the structure of the nuclei take in the cells of the reproductive organ?

The nuclei form a large proportion of each cell and are round or oval. Thread-like structures (chromosomes) may be visible.

Q2 What differences are there in the structures of the nuclei?

With good preparations it may be possible to see various chromosome stages found in gamete formation.

Male gamete (pollen) formation in Aloe

The series of photographs in figure 52 of the *Text* are presented in correct sequence and some diagrams have been made to help in their interpretation. The use of three-dimensional models to help in studies of mitosis and meiosis is recommended.

There are a number of advantages in using this sequence: *a Aloe* photographs were used in mitosis and therefore students can make a direct comparison. *b* The quality of the pictures is high enough to lead one to expect valid deductions. *c* Prophase in meiosis is not clearly seen in plant material. This is a positive advantage because crossing-over and the other details of prophase are *not* essential features of meiosis as taught at O-level.

Figure 9 contains the final sequences in this series of *Aloe* pictures. It shows the mitotic division of the haploid pollen grain nucleus to form a tube nucleus and a generative nucleus. It is unlikely that many teachers will wish to provide this extra information but the photographs are included to complete the story. Also, they will enable teachers to provide this evidence where students 'discover' that the pollen grain contains more than one nucleus.

The following notes are also by Dr P. E. Brandham.

'The meiosis series'
'**Figure 52** in the *Text* uses orcein stained pollen mother cells, squashed.
a Leptotene. Chromosomes begin to appear. Very long intertwined threads.
b Zygotene. Further condensation of chromosomes. They begin to associate in pairs (not clear).

c Pachytene. Complete association of chromosomes. A very unclear stage in *Aloe*.

d Diplotene. The eight long and six short chromosomes are now clearly visible in association in pairs as four large and three small bivalents. Each bivalent contains one or more chiasmata.

e Metaphase I. The bivalents are fully contracted and lined up on the equator of the spindle. The centromeres pull apart, but the two chromosomes in each bivalent are held together by the chiasmata.

f Early anaphase I. The chiasmata finally separate, and four large and three small chromosomes move towards each pole. In this and *h*, note that as the centromeres are near the ends of the chromosomes each chromosome appears to be two chromatids still joined near one end. Note that these two chromatids are widely divergent.

g Late anaphase I. Haploid number of 4 + 3 clearly seen at each pole. Spiralization of each chromatid is visible. The centromere which still connects the chromatids is very clear in some chromosomes in this figure.

h Telophase I – interphase. Nuclei enter a short-lived resting stage. No new cell wall is formed in *Aloe* at this point, although it is in many other plants (*e.g. Lilium*).

i–k Consecutive stages of the second prophase. It differs from a mitotic prophase in that the chromatids are still completely separated, as at anaphase I (*g*). The chromatids start off tightly coiled (again refer back to *g*). These coils gradually drop out as the chromosomes condense. Haploid number of 4 + 3 visible in *k*.

l Metaphase II. Chromatids fully condensed and completely separate, held together only by the centromeres, 4 + 3 visible in both nuclei. In one nucleus the centromeres have already separated in preparation for anaphase II.

m Anaphase II. Note that the plane of division is at right angles to the first. Full separation of each haploid set into chromatids which move towards the poles of the spindles.

n As *m* but more squashed. 4 + 3 very clearly visible in each group of chromatids.

o The intact wall of the pollen mother cell is clear, also four groups of chromatids.

p Telophase II. Each of the four groups of chromatids forms a haploid nucleus.

q Tetrad. After telophase the cytoplasm cleaves to form four potential pollen grains. Two in focus, one above focus, one below focus.

r Young pollen grain. One in each tetrad. Released by dissolving of the pollen mother cell wall. Single nucleate stage. In this stage the wall begins to become rough and sculptured.

Figure 9
A series of photographs showing mitotic division of the haploid pollen grain.
Prepared by Dr P. E. Brandham at the Jodrell Laboratory, Royal Botanic Gardens, Kew. Crown Copyright.

s A living pollen grain germinating, as on a stigma. This photograph is included to remind students of the role played by haploid gametes at fertilization.

'**Figure 9** in this *Guide* shows pollen mitosis. It is completely normal mitosis except that the chromosome number is haploid.'

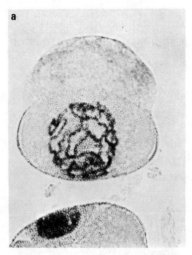

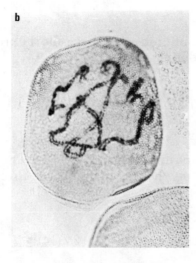

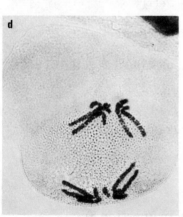

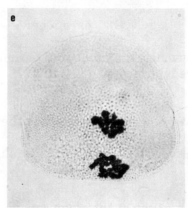

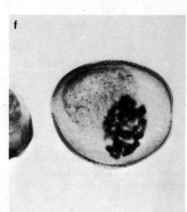

Q3 What is the normal number of chromosomes found in an *Aloe* cell?

14.

Q4 In *figure 52e* (see also the diagram in *figure 53*), the chromosomes, arranged in pairs, are about to pull apart into the daughter cells. How many pairs of chromosomes are there at this stage?

Seven pairs.

Chapter 3 The material of inheritance

Q5 Describe the way in which the chromosomes have arranged themselves in pairs.

Large or long chromosomes are paired with similarly sized and shaped chromosomes and small with small. This should give students the idea of homologous pairs.

Q6 In *figure 52g* (see also the diagram in *figure 54*), the chromosomes have separated to start forming two daughter cells. How many chromosomes are there going to be in each daughter cell?

Seven.

Q7 How many long and how many short chromosomes are going into the 'top' cell and how many into the 'bottom' cell?

Four long and three short chromosomes are going into each daughter cell.

Q8 How many threads (chromatids) has each chromosome at this stage?

Two.

Q9 Summarize the events of gamete formation seen in the sequence *a* to *h*, using the idea that the parent cell had $2n$ chromosomes.

In the parent cell $2n = 14$. The chromosomes come together in *pairs* and one chromosome from each pair enters each daughter cell.

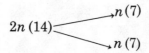

$$2n\,(14) \begin{array}{l} \nearrow n\,(7) \\ \searrow n\,(7) \end{array}$$

parent cell $\longrightarrow$ daughter cell

Q10 Why do you think that the part of the gamete formation shown in the sequence *a* to *h* is called the *reduction division*?

Because the number of chromosomes in each cell has been reduced from $2n$ (14) to n (7).

We now move on to study the 'second division' of meiosis.

Q11 How many chromosomes are going into each of the four daughter cells?

Seven (four long and three short).

Q12 How many threads (chromatids) are there now in each chromosome? Compare this answer with the one you gave to question 8.

One (two at the stage considered in question 8).

Q13 Summarize the full sequence of events from *a* to *r*. The name given to this process is *meiosis*.

Here the teacher may need to guide some students and it is an opportunity to check that the essential features of meiosis are understood.
a $2n$ number is even and chromosomes occur in pairs.
b At first division one whole chromosome of each pair enters each daughter cell (compare mitosis).
c This stage is called the 'reduction division' because each daughter cell has reduced its chromosome number to n, half that found in the parent cell, $2n$ (compare mitosis).

d At the end of the reduction division each chromosome still consists of two chromatids (compare mitosis).
e The second division is 'essentially mitotic' in that chromosome number is maintained and one chromatid from each chromosome enters each daughter cell.

Q14 *Figure 52s* shows an important stage following the sequence *a–r*. What is this stage and what is its significance in our understanding of the life-cycle?

The photograph illustrates a germinating pollen grain. It reminds us of the significance of the role of gametes in the life-cycle. Two gametes, each with *half* the number of chromosomes (*n*), contribute to the new plant with a full set of chromosomes (2*n*).

Physical models

The treatment of the concept of a model outlined in sections 2.4 and 3.2 of the *Text* can be extended by constructing physical models of mitosis and meiosis. A series of static models could be built, in which chromosomes made of Plasticine or coloured wire are held or suspended by thin plain wire. These can show the three-dimensional arrangement of the chromosomes in a cell.

Figure 10
A three-dimensional model of chromosomes.

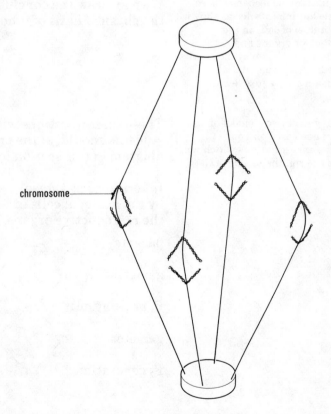

chromosome

A model which shows the movement of the cell can be constructed by making a series of drawings of meiotic figures on stiff white paper cards about 5 × 5 cm. These should be arranged in sequence and attached along one edge by staples or paper clips. By flicking the cards, the movements of the chromosomes within a cell can be simulated in quite a realistic manner. The film loop 'Meiosis' (see page 58) can be used to demonstrate the dynamic nature of the process.

3.5 Comparison with the results of the breeding investigations of Chapter 2

Objectives

1 to compare chromosome behaviour with the way factors 'behave' in the breeding investigation

2 to arrive at the logical conclusion that the chromosomes are the physical site of inheritance

The diagram in the *Text* shows the model of inheritance.

Q1 Can the results of the breeding investigations of Chapter 2 be illustrated in the same way?

Yes, in a similar way.

Q2 If similar models can be used to represent inheritance as shown by the nuclei of cells in the reproductive organs and inheritance as shown by the results of breeding experiments, what conclusion are we justified in drawing?

We may draw the conclusion that the chromosomes are the physical basis of inheritance.

Q3 Can you devise a model of inheritance based on your breeding investigations which will fit into the righthand side of the diagram?

These three questions will need discussion in class and students should, at the end, be able to complete the diagram in this section in the following way:

Inheritance as shown by the nuclei of cells in the reproductive organs

Inheritance as shown by characteristics

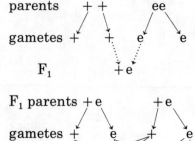

Inheritance as shown by the nuclei of cells in the reproductive organs	Inheritance as shown by characteristics
parents $2n$	parents $++$ ee
gametes n n	gametes $+$ $+$ e e
F_1 generation $2n$	F_1 $+e$
gametes n n	F_1 parents $+e$ $+e$
F_2 generation $2n$	gametes $+$ e $+$ e
	F_2 generation $++$ $+e$ $+e$ ee

Summary

It may be a good idea to ask students to summarize the work of the last two chapters. This will depend on how much you have already summarized the work in section 3.5. It will be *essential* for teachers to return to the relationship between chromosome behaviour and inheritance from time to time.

Is seeing believing?

Preformation is the idea that a miniature organism is present in the reproductive cell or gamete. All it needs is to grow and become a fully formed individual.
Ovists believed that the egg contained all the essential information for the next generation and that the semen merely stimulated development.
Spermists argued that the sperm contained a preformed individual which developed, in the egg, into a new generation.

Students may be expected to explain, in some simple way, the idea that sperm and egg both provide a complete set of haploid chromosomes which each contribute to the characteristics seen in the diploid organism. The sperm provides the stimulus to initiate development and the egg cell contains an abundant supply of food.

Background reading

Q1 Explain the meaning of the terms: preformation, ovists, spermists.

Q2 How would you describe the role of the sperm and egg in the development of a new organism?

Reference material

Books
*Reading suitable for students

*Auerbach, C. (1965) *Genetics in the atomic age.* Oliver & Boyd. (An entertaining and understandable account of meiosis is found in the early chapters.)
Barrass, R. (1964) *The locust: a guide for laboratory practical work.* Butterworth.
Biological Sciences Curriculum Study (1963) *Innovations in equipment and techniques for the biology teaching laboratory.* D. C. Heath. (Contains detailed information on keeping plants indoors.)
Darlington, C. D., and Bradshaw, A. D. (Eds) (1963) *Teaching genetics in school and university.* Oliver & Boyd. (Contains several practical articles on squash and chromosome behaviour.)
Darlington, C. D., and La Cour, L. F. (1969) *The handling of chromosomes.* 5th edition. Allen & Unwin. (Includes accounts of most of the suitable techniques for work with chromosomes.)
*Duddington, C. L. (1960) *Practical microscopy.* Pitman. (A general account of microscopic techniques including photomicrography.)
Gabriel, M. L., and Fogel, S. (Eds) (1955) *Great experiments in biology.* Prentice-Hall. (Includes excerpts from G. Newport's paper, probably the most reliable of the early accounts of observed fertilization, and L. Spallanzani's paper on experimental fertilization which includes an interesting commentary on the preformationist controversy.)
McLeish, J., and Snoad, B. (1958) *Looking at chromosomes.* Macmillan. (Contains a first-class collection of photographs of mitosis and meiosis.)

Society for Experimental Biology (1960) *Symposium XIV. Models and analogues in biology.* Cambridge University Press. (A good review of the subject.)

Articles

Shaw, G. W. (1959) 'Modern cytological techniques'. *School science review,* **41**, *143,* 88–9.
Wright, D. F. (1957) 'Models for use in the teaching of cell division'. *School science review,* **38**, *136,* 455–8.

Film loops

Nuffield Advanced Biological Science 'The use of the microscope'. Longman.
Nuffield O-level Biology 'Animal eggs and sperms'. NBP–72. Longman.
Nuffield O-level Biology 'Fertilization in the marine worm *Pomatoceros triqueter*'. NBP–71. Longman.
Nuffield O-level Biology 'Meiosis'. NBP–50. Longman.
Biological Sciences Curriculum Study 'Mitosis'. John Murray.
Nuffield O-level Biology 'Mitosis'. NBP–49. Longman.
Nuffield O-level Biology 'Squash preparation'. NBP–46. Longman.

Outline of the work on the origin of characteristics
(Chapter 4)

Introduction

4.1	The war against disease

Malarial mosquitos and DDT, rats and Warfarin, rabbits and myxomatosis

4.2	The problems of vaccination

Influenza and cold viruses, strains and mutants

Development

4.3	Mutants and mutations

Ancon sheep; albinos; haemophilia
Bacillus subtilis and antibiotic resistance (practical)
Antibiotic resistance in hospitals

4.4	Chromosomes and mutants

Chromosome aberration (*e.g.* mongolism)
Chromosomal multiplication (polyploidy in *Spartina* and clover)
Gene mutation

Background reading	How do genes work?

One gene–one enzyme hypothesis
DNA and inheritance

The origin of characteristics

1 to investigate the process of mutation in bacteria

2 to indicate the random occurrence and relatively low rate of production of mutants

3 to show that mutant organisms have arisen in many animal and plant species, including man

4 to provide evidence indicating that mutants are the result of changes in the chromosomes

5 to introduce the following concepts, while emphasizing the limited nature of the evidence provided:
a that chromosomes contain the instructions which determine the way the materials of the cytoplasm contribute to the formation of a characteristic
b that specific genes can affect particular enzymes of chemical processes in a cell (the gene–enzyme hypothesis)
c that it is possible to devise a code by which the bases of DNA could provide the instructions for the synthesis of a protein

Introduction
(Sections 4.1 and 4.2)

4.1 The war against disease

It is likely that at least some students (especially if they have done the work of Chapters 8 and 9 of *Introducing living things*) will have heard about bacteria resistant to antibiotics or insects resistant to insecticides. This, together with brief accounts similar to that in the *Text*, can be used to introduce the work of this chapter.

A display of statistics and illustrations can be constructed from materials obtainable from the World Health Organization.

It may be worth while at this stage to point out the promising 'sterile male' technique of pest control. It is based on the fact that certain insects only mate once during their life cycle and therefore millions of male insects which have been sterilized by X-rays can be released in the area *after* the insect population has been cut back by insecticide, and this will ensure the population remains at a lower level.

Myxomatosis

Q1 Suggest an explanation of the figures in the 'fatal' column.

The implied answer is that resistance has developed in the rabbit to the myxomatosis virus.

Q2 Is there another way in which the statistics could be interpreted?

The alternative is that the *virus* changed (this is known to have happened in Britain – it would, of course, be an advantage for a less lethal form of parasite if the host survived.

Q3 Design a controlled experiment to identify the correct hypothesis.

The virus could be tested against a strain of laboratory rabbits that had not been in contact with the virus.

Q4 Would you now be confident that a 'wonder drug' could be found to wipe out a disease or that a chemical could be found which would control a pest for ever?

No. The situation is a dynamic one. This question should lead the students to a discussion from which 'the problems of vaccination' should follow naturally.

4.2 The problems of vaccination

At present the mass vaccination technique seems to be the most promising. The multi-dose gun forces a small amount of vaccine under the skin using high pressure. There is no needle. Anti-viral drugs are still too toxic and at the moment it is still science fiction to consider specially engineered viruses to block normal viral multiplication in the cells of the victim.

Students are told that the next pandemic of influenza after that of 1890–91 was in the winter of 1918–19 and the 15 million deaths that it caused were mainly in the under-40 age group.

Q1 Can you suggest a reason for this?

People who were over 40 in the second pandemic and who had been exposed to the first (when they would have been at least 11) had probably caught influenza in the first. They would therefore have acquired active immunity to that strain of the disease.

Development
(Section 4.3 to end of chapter)

4.3 Mutants and mutations

The *Text* gives a number of examples of observed or apparent mutation. It is suggested that the account be augmented by a number of demonstrations. Each should consist of a wild type of the species with a number of mutants.

The sole object of this section is to show to the students

evidence of mutation as a *natural* phenomenon in a wide range of organisms.

Pupils are asked to study the family tree in *Text* figure 65.

Q1 Describe how haemophilia is inherited.

Females are either normal or 'carriers'. Males are either normal or haemophiliacs. Students may develop their description from these observations and use the idea of 'sex-linked' factors.

Q2 How did Queen Victoria become a carrier of haemophilia if none of her ancestors suffered from the disease?

The gene for haemophilia might have been 'passed down' through several generations of female ancestors who were carriers but had no sons who inherited haemophilia. More probably, the gene could have appeared as a new spontaneous mutation in Queen Victoria's mother or father during gamete formation.

Personal difficulties In dealing with human mutants it is wise to explain their features clearly and to stress their rare occurrence and, where applicable, their harmlessness. This is a precaution against students identifying themselves and others as mutants without justification and worrying about it. It is important that students realize that mutation occurs in human beings, but a well balanced picture of a rare natural phenomenon must be presented.

Care must be taken not to equate the terms mutant and abnormality. Often, to the students, abnormality implies harm, distaste, disadvantage, or eccentricity. Clearly in this sense it cannot be applied to all mutants. It is far better to talk of a mutant as a *different* characteristic rather than an abnormal one.

The distinction between mutant and mutation, too, can be confusing to the students. The idea that mutation is the *process* by which mutants arise should be established as soon as possible.

A point that may arise in discussion is the apparently disadvantageous nature of most of the mutants cited. Of course, most mutants are disadvantageous, but care should be taken to emphasize that not *all* of them are. Thus, polydactyly and syndactyly are not apparently so, and a number of examples of animals, for instance seal, bat, and frog, in which such mutant characteristics are a decided advantage, can be mentioned. So can mutant features in domesticated animals and plants that are advantageous in the artificial environment in which they live.

It is probably best not to be side-tracked too far into a discussion of adaptation and selection at this stage, but some reference to them will be unavoidable. Thus, one of the reasons why so many of the mutants detected appear disadvantageous is because they arise as new characteristics in an already well adapted organism. The chances are that few more advantageous mutants *can* be produced unless, of course, the species lives in a different environment. The fact that so many 'disadvantageous' mutants have been produced in laboratory or domesticated organisms bears this point out. In the laboratory or domesticated environment the mutants are not disadvantageous; indeed, they are selected. In nature, selection eliminates many of them.

Suggested collections for demonstrations of mutants

Living specimens are preferable but preserved ones can be used. Each collection should include the normal or wild type. Only mutants are mentioned here.

Drosophila: wing shape and size, eye colour. These can be examined with a hand lens.

Mice: coat colour, eye colour, piebald, rosette, wavy hair, belted, polydactyly, oligosyndactyly, short ear, vestigial tail, and short tail.

Lepidoptera: Melanin varieties.

Rabbits: coat colour, rex and angora.

Antirrhinum: 'giant' varieties. These are autotetraploids. Normal size varieties are diploids.

Arabidopsis thaliana:
cr (very crinkled, erect rosette leaves)
e (leaves with green veins on a yellow-green background)
le (small rosette leaves)
no (yellow leaves)
Arabidopsis is a crucifer growing to a height of 15 to 35 cm, depending on the strain. It will flower about 23 days after the seed is sown and produce viable seeds 12–15 days later. The length of the life-cycle depends on the strain, the temperature, and the light available. A temperature of 20–25 °C, and daylight supplemented by artificial light to give a total day of at least 20 hours, produce best results. *Arabidopsis* is ideal for growing in a plant enclosure (see Appendix 2, page 206). It produces abundant small seeds in a siliqua. On damp paper laid over seed compost at 25 °C, and with about 16 hours of light per day, the seeds will germinate in 2–3 days. They can be transferred to seed compost in trays. If they are sown directly onto seed compost take great care when watering them. Use a *very* fine spray.

Seeds under eight weeks old become dormant. If, however, they are first soaked in water for an hour and then kept at a temperature between 0 °C and − 10 °C for four days, they will germinate in two or three days. The seed should be collected about 16–20 days after fertilization and subjected to the cold treatment if you wish to sow them quickly. Seed over eight weeks old will germinate in two or three days at 25 °C without cold treatment.

Arabidopsis can also be used for breeding experiments but the small flower presents difficulty for an inexperienced person. *Arabidopsis* is normally self-pollinated. Cross-pollination by hand has to be carried out under a magnification of about × 10. A binocular microscope is most suitable. As the flower is slightly protogynous, emasculation and pollination can be performed at the same time.

Chrysanthemum: colouring and shape of leaves. Sports (= mutants) can be obtained from cuttings propagated from a single plant.

Cultivated cabbage: variegated and colour varieties.

Pelargonium (Geraniaceae): variegated varieties.

Primula: Primula kewensis is a known species hybrid. It was formed as a sterile hybrid from *P. verticillata* × *P. floribunda* and became fertile by a doubling of its chromosome complement. The three species grown together make a good demonstration.

Roses: climbing varieties of hybrid tea roses. Climbers are bud sports of bush roses. Incidentally, if bush and climbing varieties of hybrid tea roses are grown near together it makes a most pleasant display for the school grounds.

If quick-breeding organisms like *Drosophila* and *Arabidopsis* are kept for some time it is likely that an occasional mutation will be observed naturally. Plants like larkspur, dahlia, and chrysanthemum have genes with a sufficiently high rate of mutation to allow the occasional production of mutants to be observed. Sometimes possible mutants can be found growing wild. It is well worth while adding preserved specimens of observed mutation to the museum.

Some of the organisms mentioned above are obtainable only through research establishments. However, biological supply agencies are developing methods of cultivating many of them and should be able to supply a suitable collection.

Detecting new characteristics

Bacteria mutants resistant to antibiotics
Pathogenic bacteria should **not** be used in schools. It is
advisable, also, not to use for classwork non-pathogenic
strains of species which also have pathogenic strains,
because of the danger of producing mutant pathogens.
If adequate facilities and supervision are not available such
non-pathogenic strains could be used for demonstration.

It is a wise precaution to treat all micro-organisms as
potentially pathogenic; for contaminants can sometimes
be harmful. However, given reasonable care, the dangers
are very small, certainly no greater than those already
present in a school biology laboratory.

Students must be reminded of the importance of cleanliness
and that they should wash their hands both before and
after experimental work, not put their fingers, pencils, etc.,
into their mouths, and obey the rules of sterile technique
conscientiously.

Mutation can be studied in bacteria. *Bacillus subtilis* is
harmless and recommended for this work. However, it
quickly forms large quantities of spores and if exposed
unduly in a laboratory can be an annoying source of
contamination for a long time. The spores are very
resistant to dry heat but autoclaving is usually an adequate
method of control.

Teachers may find it useful to refer to the notes on the
culture of bacteria, including necessary precautions,
which are also contained in *Text* and *Teachers' guide 1,
Introducing living things,* in Chapters 7 and 8.

Culture media
For notes on these, see Appendix 1, page 194.

Antibiotic discs
The most effective for a particular strain of bacteria can
only be determined by experiment so it is recommended
that several types of sensitivity disc be used on each Petri
dish, for example, penicillin G, streptomycin, tetracycline,
chloramphenicol. The alternative is to use the Multodisk.
This is a paper disc with eight arms, each tip impregnated
with an antibiotic. 30–2–G and 30–2–H are economical to
use. They have four antibiotics, penicillin G, streptomycin,
tetracycline, and chloramphenicol and each disc can be
cut in half (use *sterile* forceps and scissors). (These discs
can be obtained from Oxoid Ltd.)

Bacteria cultures
These can be purchased from biological supply companies or from the National Collection of Type Cultures, Central Public Health Laboratory, Colindale Avenue, London NW9. *Bacillus subtilis* is suitable for this investigation, also *B. cereus*, and the coloured bacteria, *Serratia marcescens* (red) and *Sarcina lutea* (yellow), but the last two should be incubated at 20 °C.

The culture will be delivered growing on an agar slope. Pour some nutrient broth into the tube and shake it. Then add this suspension to the nutrient broth in a conical flask and incubate at 37 °C for 24 hours. Use a sterile plastic disposable syringe to dispense 2 cm^3 portions of this culture into sterile tubes (screw-topped McCartney bottles).

Apparatus and materials
Per class:
2 tubes sterile nutrient agar (15 cm^3) (see Appendix 1)
McCartney bottle with 2 cm^3 of bacteria culture
sterile McCartney bottle with 2–3 cm^3 nutrient broth (see Appendix 1)
autoclave (or pressure cooker)
incubator at 37 °C
water bath at 45–50 °C
bowls of disinfectant (combined disinfectant and detergent recommended for swabbing benches; Chapter 7 and Appendix 1 of *Text 1* give further details)
wire inoculating loop (nichrome wire 10 cm of 26 or 28 s.w.g.)
2 sterile Petri dishes
Oxoid Multodisk or separate sensitivity discs
Chinagraph pencil
Sellotape
Bunsen burner
gauze
tripod
beaker

Figure 11
A wire loop used for transferring micro-organisms.

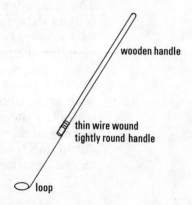

wooden handle

thin wire wound
tightly round handle

loop

The plates should be inspected after incubation for 24 hours and if sufficient growth has been made they can be held in the refrigerator until the next practical lesson.

Q3 What do your results indicate?

Q4 Is there still a zone of inhibition around the antibiotic disc?

Q5 Are there more resistant colonies?

One must suppose the original strain to be homogeneous, that is, to be derived from a single spore sensitive to the antibiotic. Mutation to antibiotic-resistant strain could therefore have occurred in the rapid division under optimum conditions in the nutrient broth or on the agar plate.

Q6 When do you suppose the new characteristic developed?

Q7 Has the presence of the antibiotic caused the bacteria to mutate?

No! The mutation was spontaneous, the antibiotic was merely acting as a factor in selection.

Q8 Why is it important to use non-pathogenic bacteria in this experiment?

It could be dangerous to produce a strain of pathogens resistant to attack by the more common antibiotics.

Q9 Why is it important to dispose of the plates by autoclaving even though *B. subtilis* is non-pathogenic?

The plates could be contaminated by other species of bacteria, some of which could be harmful, or a new mutant *might* appear.

Q10 Suggest an explanation for the results in *table 9*.

A mutant bacterium which was resistant to penicillin could survive the effects of penicillin. This mutant strain became the most common variety of *S. aureus* in the hospital concerned, over two years, as penicillin killed other strains of *S. aureus*.

Isolation of streptomycin-resistant bacteria
This can also be demonstrated, using a more sophisticated method – the 'gradient plate' technique.

Preparation of 'gradient plate'
1 Tilt a sterile Petri dish on the edge of a ruler and pour into it a tube of molten nutrient agar. Allow the agar to set in this position.
2 When it has set, place the dish flat on the bench and pour into it a tube of molten streptomycin agar. Allow it to solidify. Such plates will have a streptomycin concentration gradient proportional to the thickness of the streptomycin agar.

Figure 12
A 'gradient plate' for the isolation of bacteria resistant to streptomycin.

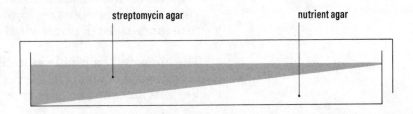

streptomycin agar   nutrient agar

Isolation of resistant mutants

1 Add about 0.5 cm^3 of *Escherichia coli* suspension to the plate and, using a sterile glass spreader, inoculate the whole surface of the agar. Incubate at 30 °C until the next laboratory period.

2 Look for resistant colonies. Select two or three colonies which are growing at the high streptomycin concentration and streak these in the direction of the highest streptomycin concentration. Re-incubate at 30 °C until the next period.

3 With a wire loop, prepare a suspension in nutrient broth from two or three colonies growing at the highest streptomycin concentration. Use this broth suspension to inoculate tubes of nutrient broth containing 0, 0.1, and 0.5 mg cm^{-3} of streptomycin. Inoculate similar series of broths with the cultures provided. Incubate at 30 °C. Record growth during the next period.

Here are two other practical investigations which the class might undertake.

1 Detecting resistance to insecticide in houseflies and *Drosophila*

Houseflies are easily obtained and are of obvious social importance. They are good material for experimental work. To detect resistance to insecticide a large number should be caught, using a few lumps of sugar as a bait.

Between 25 and 50 flies should be put in an inverted beaker or gas jar containing insecticide of known strength. The number of flies killed after one hour can then be recorded. This procedure will give a rough criterion for resistance.

More defined work can also be undertaken, such as:

Determining resistance to different concentrations of insecticide. The numbers killed during a given time in different concentrations of insecticides can be graphed.

Determining the influence of exposure time on resistance. Count the number of flies killed in a concentration of insecticide at given time intervals, for example, every 15 minutes. Graph the results.

Resistant flies can be bred and crossed and an attempt made to elucidate the genetic nature of resistance. The genetics of insecticide resistance in the housefly are not yet fully understood. It is known that in some forms, resistance to DDT is inherited as a single recessive gene.

The housefly is a highly suitable organism for long-term investigations and project work by individual pupils. Various mutants, besides those with resistance to insecticide, can sometimes be obtained from research establishments.

Other insects, including *Drosophila*, can be used to detect mutants resistant to insecticide. The wild *Drosophila* can be used for this work. It can be caught by using an over-ripe banana, pear, orange, or similar fruit as a bait, and kept in milk bottles with a cottonwool bung and fed on ripe fruit.

2 Mutation in yeast

Haploid *ad* yeast is a red mutant form from the wild white type. It will readily undergo mutation to a white form and even after one day's incubation, white colonies are observed. They quickly outgrow and obscure the red so that, although under ideal conditions the experiment is a very convincing demonstration of mutation, it is unreliable in the school laboratory.

If a demonstration is planned:
1 medium and *ad* strain can be purchased from laboratory suppliers
2 keep in the refrigerator until required
3 be careful to inoculate plates only from red colonies
4 incubate at 28–30 °C for about two days but inspect frequently and transfer to refrigerator as soon as any white colonies appear
5 incubate an unopened plate as control.

Mutation rate

It is important for students to realize that mutation is a *natural* phenomenon but most of them will know the mutagenic effects of ionizing radiation. Müller showed that the frequency of mutation is directly proportional to the dose of X-rays received.

4.4 Chromosomes and mutants

Once the students have accepted mutation as a natural occurrence they should examine evidence of the genetic origin of mutation. A comparison of the chromosomes of normal and mutant organisms is an appropriate way of doing this with fifth-form students. The main aim of the work is to show that the presence of mutant characteristics can be related to change in the structure of the chromosomes.

Three types of mutation can be considered:
1 *Chromosome aberration*, in which a whole, or a part of a chromosome becomes rearranged. This covers non-disjunction, translocation, deletion, duplication, inversion,

etc., but, of course, there is no need to consider any of them separately. Mongolism is the main example used. The failure of the ovary to develop (ovarian dysgenesis) is another. Both are due to non-disjunction – the failure of two members of a chromosome pair to separate at metaphase. However, it is only necessary for the students to realize the way the chromosome complement differs from the normal in each case, namely, the extra chromosome of the 21 group in a regular mongol; the lack, or reduction in size, of an X chromosome in the case of ovarian dysgenesis.

Associated with ovarian dysgenesis is the failure of the secondary sexual characteristics to develop. Thus, the difficulty of diagnosing the cause of certain characteristics can be pointed out to students. No doubt they will already have learned of the influence of the endocrine glands on sexual development. Now they can see that it has a genetic basis also. The inheritance of sex could be discussed further at this stage.

Q1 How are the chromosomes of a mongol different from those of a normal person (compare *figures 70* and *71*)?

The mongol has an extra chromosome compared with a normal human being. As some mongols would appear to have been produced by mutation it is a reasonable hypothesis to suggest that the production of the extra chromosome was the cause.

Figure 72 in the *Text* shows the chromosomes of a woman whose ovaries did not develop properly and who lacked secondary sex characteristics.

Q2 Do you consider that this is another case of mutation due to chromosome aberration?

The characteristics of the sexually under-developed woman are correlated with the presence of a small chromosome. As such people occur very rarely it is likely to be a case of mutation. Hence this example is similar to that of the mongol.

Q3 Does it add further support to the hypothesis that has been suggested?

Harris Biological provide photographic sets for use with this work (see the Reference material, page 77).

2 *Chromosome multiplication*, in which the number of chromosomes is increased as a multiple of the haploid number. This covers all types of polyploidy but, again, there is no need to go into details.

A possible complication in discussions about chromosome multiplication is the idea that there is a relationship between the number of chromosomes and the number of characteristics in an organism. This is not so. For example, in man $2n = 46$, in the crayfish $2n = 208$, and a radiolarian (a protozoan) with about 13 600 chromosomes has been reported.

Q4 What do you think is the explanation of the origin of *Spartina townsendii*?

Huskins considered that originally plants of *Spartina maritima* ($2n = 60$) and *Spartina alterniflora* ($2n = 62$) crossed, but in the process, doubling of the numbers of chromosomes occurred, so that *Spartina townsendii* was produced, containing the sum of the diploid number of chromosomes of the original species ($2n = 122$).

3 *Gene mutation*, in which changes occur in the chromosomes at the molecular level. This poses a problem, for it is impossible to observe the changes as one can observe chromosome aberration or multiplication. It is suggested that the topic be presented to students as a question. Mutants such as haemophilia, and ebony body, appear in a population, just as mongols and the *Spartina* mutants do, yet no change of number or form of their chromosomes can be observed. Is it possible that a change in the chromosomes may be responsible for them?
As a hypothesis it is feasible for students to consider that an unobserved change at the molecular level could be responsible. Indeed an alternative explanation is difficult to conceive. It is best left as hypothesis, for research into the nature of gene mutation is still largely at the hypothetical stage and a detailed consideration of the subject would be inappropriate with fifth-form students. The majority of mutations are gene mutations. This point should be made clear to the students.

One should aim to get the kinds of answers shown below to the questions put to the students at the end of this section.

Q5 We have given a name to the process by which these characteristics arise – mutation. What conclusion can you draw from your studies about the nature of mutation?

Mutations occur in all kinds of living organisms. They result in new characteristics in organisms. Once established, a mutant characteristic is inherited. Mutations occur at definite rates but it is not possible to predict when one will actually occur; in this respect they are random.

Q6 What comparisons can be made between the processes of mutation and inheritance?

Mutation produces a new characteristic. Inheritance transfers the same characteristics from one generation to another. Mutation produces a characteristic which is very rare, while inherited characteristics occur in greater numbers. A characteristic produced by mutation is inherited.

Q7 Do the chromosomes give any clue to the nature of mutation?

Visible changes in chromosomes, that is, chromosome aberration and multiplication, can be related to the production of new characteristics. Most cases of mutation occur without visible change in the chromosomes, but it is likely that they are due to changes in the chromosomes that cannot be observed with present-day techniques, that is, to gene mutation.

Q8 Can you construct a general
hypothesis to explain how
inherited characteristics arise?

Q9 How would you test it?

Inherited characteristics arise by the process of mutation
in which changes occur in the chromosomes of a cell.

Devise a method of altering chromosomes without
influencing any other part of the cell, so that you can show
directly that a change in the chromosome can result in a
mutant characteristic.
Experiments with colchicine, X-rays, and similar agencies
could be cited to illustrate this type of investigation. Of
course the students would not be expected to work out how
this could be done practically.

Human chromosomes

The first significant count of human chromosomes was
made by von Winiwater in 1912. He considered 47 to be the
diploid number in the male. However, until 1956 there was
a great deal of uncertainty about the actual number.
Counts were made on sectioned material obtained mainly
from people who had recently died but the technique led to
distortion, and although most workers accepted 48 as the
diploid number with an XY chromosome relationship in
the male, it was still open to doubt.

In 1956 Tyco and Levan in Sweden, and Ford and Hamerton
in England, reported 46 as the diploid number. Using squash
techniques they had separated out the chromosomes
sufficiently to make extremely accurate counts. The
photographs in the *Text* are from such preparations. The
outline of the technique is as follows.

White blood cells are cultured and induced to divide by
phyto-haemaglutinin obtained from kidney beans.
Colchicine, or some other mitotic poison, is used to disrupt
mitosis so that the chromosomes are scattered in the cell.
A hypotonic fluid is placed around the cells which become
swollen. The cells are collected from the culture by
centrifuging, which also helps to scatter the chromosomes
further. The cells are placed on a slide, squashed, and
stained, and can then be easily examined.

The accepted diploid number of human chromosomes is
now 46. The chromosome pairs are numbered 1–22 in order
of decreasing size according to the Denver system, a
standard system proposed by a symposium at Denver,
Colorado, in 1959 and now universally accepted. The X and
Y chromosomes are labelled separately. Identification is
not always easy and comparison with photographs
identified by an expert is necessary for the beginner. The
photographs in the *Text* can be used as approximate
standards.

Preparations of human chromosomes could be produced at school by a teacher, a technician, or able and interested students. Details of the techniques can be found in Hamerton (1965), and in Darlington and Bradshaw (1963).

Although the vast majority of human cells will have a chromosome complement of 46, variations are occasionally found. An example of the range normally encountered in a sample of 100 cultured white blood cells is:

Number of chromosomes	<44	44	45	46	47	48	>48	Total
Number of cells		1	2	2	93	2	—	—	100

Polyploidy has been reported in human beings. For example polyploid cells have been found in the liver and other organs. Also a triploid individual – a deformed boy – has been identified.

Suggestions for homework or preparation

The problems included in the *Text*, which do not require practical facilities, could be undertaken for homework or preparation.

Other problems

1 A dwarf bull calf is born to apparently normal parents in a herd of cattle. How might you decide whether the calf is the result of *a* mutation, *b* the mating of two carriers of a recessive gene for dwarfness, *c* environmental influences?

2 Haemophiliacs are less likely to achieve a normal life span than normal humans, and the rate of elimination of haemophilia-causing genes by death of the persons bearing them is between 1 and 5 in 10^5 per generation. If the number of haemophiliacs in the human population is constant over the years, what is the rate of mutation of the normal gene to the haemophilia-causing gene?

3 In two families the characteristics of short fingers and albinism had never been recorded previously. However, in one, a short-fingered boy was born who later married a normal woman and had three normal children and one with short fingers. This short-fingered offspring married a normal woman and they had three children all with fingers of normal length.
In the other family, an albino child was born. He married a normal woman and they had four children, all normal. What are the most likely explanations for the appearance and disappearance of these characteristics in each pedigree?
(The characteristic of very short fingers – brachydactyly – in human beings has been shown to be due to a single dominant gene. Albinism is due to a single recessive gene.)

How do genes work?

Students have arrived at the point in their studies where they know something about the chromosomes and something about the pattern of inheritance of characteristics. They should realize also that there is an apparent relationship between them. Now it is necessary to indicate what this relationship is.

The chromosome → cytoplasm → characteristic model and previous work in the course suggest that the connection between chromosomes and characteristics lies in the cytoplasm of the cells. It follows from this that there should be a connection between the chemical processes in the cytoplasm of the cell and the genes in the chromosomes.

The experiments of Beadle and Tatum with *Neurospora* lead to the concept of a one to one relationship between genes and enzymes and also between a mutation and a block in a metabolic pathway.

Q1 What is the difference in growth between strains 1 and 3?

Arginine is needed for growth. Strain 6 is unable to manufacture arginine so it must be supplied. There is no point in adding ornithine to strain 3; it is unable to convert it.

Q2 What is the difference between strain 3 and strain 6?

Q3 What could account for these differences?

Strain 1 is unable to manufacture ornithine so it grows if ornithine is supplied and also if the early stages are bypassed and citrulline or arginine is supplied.

Q4 Do these results agree with the proposed pathway?

Yes, the sequence is confirmed as a series of chemical stages. Mutant 1 is an ornithine-requiring strain, mutant 3 citrulline-requiring, and mutant 6 an arginine-requiring strain.

Q5 What effect do you think that genes might have in controlling this metabolic pathway?

A different gene is responsible for each stage in the chain of reactions. As it is known that the chemical reaction of each stage is regulated by an enzyme it is likely that genes control the production of the enzyme. If the gene is changed the enzyme is unable to function.

The metabolic pathway is shown in *figure 13*.

DNA and inheritance

The aim of this part of the Background reading is not to consider the detailed structure of DNA but to provide an outline picture showing that it is made up of a series of nucleotides which differ only by the nature of their bases.

Four concepts are considered:

1 The concept of instruction
2 That it is the bases of the nucleotides that act as instructions.
3 That combinations of the four bases could provide the necessary instructions responsible for an organism's characteristics.
4 That it is possible to devise a scheme by which combinations of the bases in triplet can provide the instructions for building up a protein molecule.

Figure 13
From Beadle, G. W. (1946) 'Genes and the chemistry of the organism', American scientist **34**, *131–53.*

It is best to keep strictly to the elementary outline described in the *Text*. However, in discussion the following points may be useful.

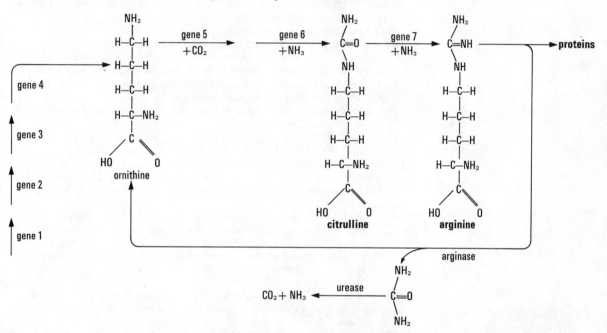

The argument for the triplet code rests on the grounds that one of four single bases, or pairs of bases, would be insufficient to code 20 amino acids. A triplet code allows 64 combinations, which would be sufficient. A quartet code would provide far too many superfluous combinations to be credible. Experimental evidence now supports the idea of a triplet code. It also indicates that it is non-overlapping; that is, the coding of an amino acid is performed by a triplet that has no bases in common with the next triplets. The triplet code is also assumed to be degenerate; that is, several triplets can code the same amino acid.

It is suggested that details of RNA, messenger RNA, and the triplet code are unnecessary at this level. The enquiring student can be led to more advanced texts.

Summary

At this point, students should be asked to make their own summary of the work of this chapter.

Reference material

Books

*Reading suitable for students

*Alexander, P. (1957) *Atomic radiation and life.* Penguin. (Students will find some parts difficult.)
*Ashton, B. G. (1967) *Genes, chromosomes and evolution.* Longman.
*Auerbach, C. (1965) *Genetics in the atomic age.* Oliver & Boyd.
Biological Sciences Curriculum Study (1963) *Innovations in equipment and techniques for the biology teaching laboratory.* D. C. Heath. (Contains useful suggestions about apparatus for microbiology.)
*Biological Sciences Curriculum Study (1963) *Molecules to man.* Arnold. (Views the whole of biology from a basis of molecular biology. Excellent sections on genetics.)
Book of life (1971) (A 105-part magazine.) Marshall Cavendish.
Darlington, C. D., and Bradshaw, A. D. (Eds) (1963) *Teaching genetics in school and university.* Oliver & Boyd. (See Bevan, E. A., and Woods, R. A. 'Yeast in practical genetics', 29–35; and Ockey, C. G. 'Peripheral blood cultures of human chromosomes', 64–7.
The Difco Manual (1960) Baird & Tatlock Ltd, Freshwater Road, Chadwell Heath, Essex.
Fincham, J. R. S., and Day, P. R. (1963) *Fungal genetics.* Blackwell.
Garbutt, J. W. (1972) Students' Manual and Teachers' Guide, *Experimental biology with micro-organisms.* Butterworth.
Hamerton, J. L. (Ed.) (1965) *Chromosomes in medicine.* Heinemann. (A full account of human chromosomes and their relation to clinical abnormalities.)
*Kelly, P. J. (1962) *Evolution and its implications.* Mills & Boon. (Contains an elementary account of the genetic effects of radiation.)
Levine, R. P. (1962) *Genetics.* Holt, Rinehart & Winston. (Contains a good, concise account of the genetic control of biosynthetic pathways.)
*Nuffield Biology, Revised edition (1974) *Text* 1 *Introducing living things.* Longman. (See Chapter 7, 8 and 9.)
Nuffield Biology, Revised edition (1974) *Teachers' Guide* 1 *Introducing living things.* Longman. (See Chapters 7, 8, and 9 and Appendices 1, 2, and 3.)
The Oxoid manual (1971) Oxoid Ltd, Southwark Bridge Road, London SE1 9HF.
Peters, J. A. (Ed.) (1959) *Classic papers in genetics.* Prentice-Hall. (Contains original papers of many of the significant discoveries in genetics including ones dealing with the structure of DNA and bacterial transformation. With guidance, able and interested students would find parts of this book useful.)
Report on Public Health No. 100 'Influenza epidemic in England and Wales 1957–58'.
Stern, C. (1960) *Principles of human genetics.* W. H. Freeman. (A wide and, at times, advanced treatment with a good chapter on the genetic hazards of radiation.)
Wald, G., *et al.* (1967) *Twenty-six afternoons of biology.* Addison-Wesley. (Contains information on practical techniques for bacterial transformation.)
World Health Organization (June 1957) *Assessment of susceptibility to insecticide in anopheline mosquitos.* World Health Organization.

Articles

*Reading suitable for students

Busvine, J. R. (1958) 'Housefly colonies for school biology'. *Journal of the Institute of Biology*, **6**, *1*.

Busvine, J. R. (1961) 'Insecticide-resistant strains of insects in England by 1961'. *The Sanitarian*. December 1961, **70**, *3*, 192–3.

Crick, F. H. C. (1962) 'The genetic code'. *Scientific American* Offprint No. 123.

Crick, F. H. C. (1954) 'The structure of the hereditary material'. *Scientific American* Offprint No. 5. (Like all the *Scientific American* Offprints listed here, this provides a good account of its subject. Able and interested students would find these offprints useful.)

Fagle, D. L. (1961) 'Bacteriology course'. *The science teacher* **28**, *7*, 24–7. (Contains good, simple ideas for practical sterile technique in the school laboratory.)

Hoagland, M. B. (1959) 'Nucleic acids and proteins'. *Scientific American* Offprint No. 68.

Hotchkiss, R. D., and Weiss, E. (1956) 'Transformed bacteria'. *Scientific American* Offprint No. 18.

Lambert, J. M. (1964) 'The *Spartina* Story'. *Nature*, **204**, 1136–8.

*Mangelsdorf, P. C. (1953) 'Wheat'. *Scientific American* Offprint No. 25. (A good account of the origin of modern bread wheat.)

Riley, R. (1959) 'Chromosomes and wheat breeding'. *Times science review*, Autumn 1959. (Places emphasis on the genetics of wheat evolution.)

Society of American Bacteriologists Committee of Education (1960) 'Microbiology in introductory biology'. *The American biology teacher*, 22(6). (Contains much information on work with micro-organisms in schools and particularly useful practical hints. Mainly concerned with bacteriology.)

Sorsby, A. (1958) 'Noah – an albino'. *British medical Journal*, 5112, 1587–9.

Films

'Genetics and plant breeding', sound, colour, 17 minutes. 16 and 35 mm. Unilever Films, Unilever Film Library, Unilever House, London EC4P 4BQ.

'Chemistry of the cell', sound, colour, 16 mm. Part I 'The structure of proteins and nucleic acid', 22 minutes. 601 D28. Part II 'The function of DNA and RNA in protein synthesis', $16\frac{1}{2}$ minutes. 601 D29. Distributed by E.F.V.A. Paxton Place, Gipsy Rd, London SE27.

Photographs

Human chromosome analysis sets, Harris Biological Supplies Ltd, Oldmixon, Weston-super-Mare, Somerset.

Outline of the work on development
(Chapters 5 and 6)

5.1   **Nucleus and cytoplasm**

Acetabularia Xenopus and chimeras

5.2   **Inheritance and environment**

5.21   Do inheritance and the environment together determine an organism's characteristics?

Albino seedlings in light and dark: 'nature and nurture'

5.3   **The influence of chemicals on development**

Plant hormones and seedlings

5.4   **Animal hormones and development**

Castration; metamorphosis

5.5   **Measuring variation in development**

In boys and girls; principles of measuring variation

Background reading   Unofficial life

Development of embryo

6.1   **Different patterns of development**

Life cycles

6.2   **Is cell division related to the pattern of development?**

Graphs of barley

6.3   **How do cells differentiate?**

Root cells

6.4   **Interaction between cells**

6.5   **Control of development in the whole organism**

Regeneration in *Planaria* and *Begonia*

Background reading   Carrots, coconuts, and cancer

Regeneration in single cells

5
Development

An alternative sequence for the work in Chapters 5 and 6

1 Start with variation in human development and the problems of measuring variation — *section* 5.5

2 Study the Background reading for Chapter 5, 'Unofficial life'

3 Work through Chapter 6 in sequence — *sections* 6.1–6.5

4 Pose the problems of the role of nucleus and cytoplasm — *section* 5.1

5 Investigate the factors influencing development — *sections* 5.2–5.4

Before starting the work of these two chapters, you might present the students with a large diagram to show the route you intend to take.

Advanced preparation (Chapter 5)

5.4 If you intend to do practical work on metamorphosis in amphibians, a supply of tadpoles (not less than 50) will be required. They should be aged about 6–10 weeks depending on feeding and temperature of culture (see 'Amphibia' in Appendix 2, page 197).

6.3 Onion roots must be collected in time for the practical work. These can be grown in advance and stored in fixative (see 'Roots' in Appendix 2, page 219).

Growth and development

Throughout the work of Chapters 5 and 6 the term *development* is used exclusively, in order to avoid confusion over the definition of the terms growth and development.

A definition of development as *the changes involved in producing a mature (= fully developed) organism* covers the vast majority of definitions of growth and development. It makes the use of the word growth unnecessary.

Some biologists define growth as an increase in the living material of an organism. Others will say it also includes changes in shape, appearance, and function. Some biologists mean by growth those aspects of the changes a living organism undergoes, from the egg to the adult stage, *which can be measured*. For them, development is those aspects that *cannot be measured*.

This apparent confusion over the definitions of growth and development is a convenient focus for a discussion on the importance of both clearly defining terms and not adopting superfluous ones. It could be extended to a consideration of the use of terms in scientific and non-scientific writing. Terms like development and inheritance are frequently used by writers for different purposes. Their use in everyday speech is also variable. It would be a useful exercise for the students to make a collection of the different meanings given to them. This could provide a useful link with their English studies.

5.1 Nucleus and cytoplasm

Objectives

To examine some experimental evidence indicating the influence of the nucleus and cytoplasm on development and to note some of the problems that arise

In Chapters 3 and 4 the students' work was focused on the nucleus as the possible source of an organism's characteristics. Reservations were made at the time and they will make a good introduction to the work of Chapters 5 and 6.

Q1 If a fertilized human egg splits apart at the two-cell stage, and the daughter cells continue to develop, what is the result?

Identical twins.

Figure 14 could be used at this point to show the difference between identical and non-identical twins.

Q2 The nine-banded armadillo always has quadruplets which are of the same sex and show no obvious variation. What does this indicate?

They are likely to be identical quadruplets and derived from a single fertilized egg which separates at the four-cell stage.

From the results of these questions a series of problems is raised in the *Text*.

An experiment with *Acetabularia*

The experiment with *Acetabularia* described in the *Text* is one of a series undertaken by Professor J. Hammerling published in 1943. The series is referred to in many textbooks, for example, Srb and Owen (1965), Ebert (1970); see the Reference material on page 104.

Q3 What do these results suggest about the respective roles of the nucleus and cytoplasm in development?

They indicate that the nucleus influences the development of the cytoplasm. Thus the new 'hats' developed under the influence of the bases, not the stems. If we assume that the difference between the stem and the base is solely the presence of a nucleus in the base and its absence in the stem, it is reasonable to conclude that the nucleus plays the predominant role in the development of this organism.

The perpetuation of life

Figure 14
Diagrams to show the origin of
a non-identical twins, when each
ovary releases an egg and both are
fertilized
b identical twins, when one egg is
released and fertilized, and divides
in two.

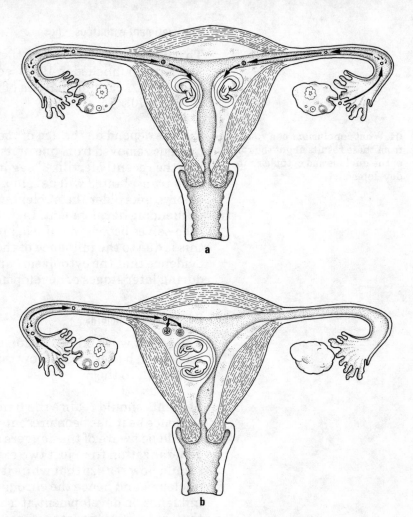

a

b

The film loop 'Removing and exchanging nuclei in
Amoeba' (see the Reference material, page 104)
illustrates the importance of the nucleus to the life of the
animal. Details of its contents and suggestions for its use
will be found in the teaching notes accompanying it.
It can provide a good supplement to the work on
Acetabularia.

The film loop 'Regeneration in *Acetabularia*' (see page 104)
is also useful.

An experiment with frogs' eggs

This description is based on the investigations of Briggs and King, published in 1957 (see the Reference material, page 104). A careful analysis of this work is to be found in Ebert (1970).

Q4 What conclusions can you draw from these results about the role of the nucleus and cytoplasm in development?

Results depend on the age of the embryo which had the nucleus removed from one of its cells. The older it was, the less the recently fertilized egg into which the nucleus has been transplanted will develop. It seems that as an embryo gets older the nuclei tend to lose their power of influencing development. In other words the nuclei themselves become modified, and it may be assumed that this is due to the influence of the cytoplasm. Thus there is evidence that the cytoplasm can influence the nucleus during later stages of development.

Human–chick chimeras

The recent work on cell fusion is introduced here to give students a better idea of how this sort of work is important to them.

Students should realize the limitations of this experimental evidence as it has been presented to them. Clearly they should be aware of the dangers of making too sweeping a generalization from just two experiments. However, they should now realize that while it is logical to assume that the nucleus – and hence the chromosomes – has a predominant influence on development, it is also reasonable to consider that, at least in the later stages of the development in multi-cellular organisms, the cytoplasm can, to some extent, influence the nucleus.

5.2 Inheritance and environment

Objective

To provide a framework within which students can attempt to answer experimentally the question: 'In what way, if at all, do inheritance and the environment determine an organism's characteristics?'

The example of the influence of nutrition on the body proportions of pigs is used to provide an introduction to the problem of distinguishing the influence of inheritance and the environment on development. It can possibly be made more credible to students by indicating the influence that nutrition can have on the body form of human beings.

The *Text* describes how pigs were given the Chinese quad treatment.

Q1 Why is this term used?

The term refers to the different types of diet (planes of nutrition) during the life of the animal: hi-hi, lo-hi, hi-lo, and lo-lo.

The perpetuation of life

The economics of different diets is very relevant here in relation to agricultural practice. Farmers who rear beef and dairy cattle tend to raise the calves together. In fact the dairy calves should be reared on a low–high programme: low diet at first so that the animal does not develop too big a frame and then a high plane of nutrition during pregnancies and lactation.

Beef cattle should be reared on a high–low programme: high diet at first to develop a large frame, and low plane of nutrition during later development to prevent excessive fatness.

Twin studies

Comparative studies of twins and siblings provide data illustrating the inter-relationship of inheritance and the environment. Identical twins produced as a result of the separation of cells at the first division of a fertilized egg can be assumed to have largely the same genetical constitution. Non-identical twins and brothers and sisters (siblings) can be assumed to have genotypes showing much greater differences. Thus, on genetical grounds, one would expect greater differences between the characteristics of siblings and non-identical twins than between those of identical twins. Research has shown that such differences exist, but that it is difficult to determine how much they are a result of environmental influences.

Comparative studies of sets of identical twins show that there are greater differences between members of sets of twins reared apart than between those reared together. Thus, there is evidence that environmental influences affect the expression of the genotypes.

Asking the right question

This section emphasizes the importance of clearly formulating problems. A problem is the starting point of any investigation, and the way in which it is formulated dictates the direction an investigation will take.

In order that a problem can have meaning from the point of view of enquiry it must be formulated as a question and it is this question that is so important in determining the future path of enquiry.

It is worth while bringing this point home to students because it helps to make them conscious that the methods of thinking they employ can influence investigation. It makes them realize the subjective element in their work.

For 15-year-old students it is wise not to make topics such as this appear too abstract. They should be linked with specific examples, as in this section. Once the point has been established it can be brought up whenever questions are posed for investigation. The question, 'Am I asking the right type of question?', should become the immediate reaction of a student when a problem is being tackled.

5.21 Do inheritance and the environment together determine an organism's characteristics?

It is worth noting that contamination is a potential danger in all experiments on development. It is important to refer students to the sterile techniques used in their work with micro-organisms and to point out that the same principles should be applied to their present work.

Apparatus and materials
Per group:
100 seeds (approximately)
2 Petri dishes or seed trays
filter paper, or agar medium, or seed compost
black paper to cover a Petri dish
adhesive tape
plant labels

Experiment 1

The seeds given to the students for sowing will be from F_1 heterozygotes. Green is dominant to albino, except in tomato where it is incompletely dominant. It is not necessary for each group of students to sow 100 seeds but at least that number should be sown by a class.

Selfing the F_1 yields an F_2 of 3 green: 1 albino.
Thus, the students should be able to deduce from the ratio of the colour types observed in the F_2 and the colour type of the F_1 which they are told about in the *Text*, that the characteristics of green and albino follow a pattern of inheritance similar to that which they have already studied.

Q2 What percentage has germinated?

Usually well over 80 per cent. If it is lower than 60 per cent the obtained proportion of seedlings should be compared with the expected ratio to see if one characteristic is being affected more than others.

Q3 How many seedlings of each colour have developed?

Sample answers are:
Tobacco 75 green: 19 white
Tomato 24 green: 47 pale green: 18 white

Q4 If your results suggest that the characteristics of green colour and albino are inherited, work out the pattern of the inheritance of the characteristics. What genes responsible for colouring are carried by the seedlings that develop? What genes do you think their parents carried?

It is a case of monohybrid inheritance similar to those studied in Chapter 2. It can be shown in outline as follows, using these symbols for convenience:

G = dominant gene for greenness

g = recessive gene for albino

F_1 Gg × Gg   **all green**

F_2 $\underbrace{\text{GG Gg Gg}}$, gg

          3 green : 1 white

In the case of tomato:

G = gene for greenness

g = gene for albino

F_1 Gg × Gg   **pale green**

F_2 GG $\underbrace{\text{Gg Gg}}$ gg

       1 green   2 pale green   1 white

Q5 Can you devise an experiment that would check your conclusions?

Self members of the F_2 generation to confirm their genotypes. (Time and the nature of the albino plants, which the next experiment should show, would make this difficult to do as part of the course.)

Experiment 2

This is best performed by sowing about the same number of seeds as in experiment 1, but covering their containers with a matt black paper and, if possible, placing them in a dark cupboard.

Under these conditions all the seedlings should become albino. Experiment 1 (in the light) and experiment 2 (in the dark) act as controls to each other. This is necessary for the interpretation of the results of experiment 2.

Experiment 3

Q6 What conclusions do you draw?

A combination of experiments 1 and 2 confirms the hypothesis that inheritance and environment together determine an organism's characteristics. From the experiments it would appear that in the case of green colour, inheritance and the environment are jointly responsible. The gene has to be present in the organism, but its *expression* depends on the nature of the environment; that is, in this instance, light has to be present.

In the case of albino plants it would not appear that the environment of either experiment has any influence. However, the experimental environments are limited and it is important to bear in mind the proviso that under natural conditions other factors may be influential. For example, certain inorganic salts can influence the colouring of plants.

Barley (Hordeum vulgaris)   Seeds are sown in a tray of moist sand. Segregation is discernible after about ten days at 20–25 °C.

The albino seedlings will die eventually. They are incapable of photosynthesizing. Thus, in order to maintain a strain of these plants, the following procedure must be adopted:

Transplant a number of the green seedlings into trays of soil. When the leaves are about 5 cm long set the plants out in individual pots.

Bag the flower buds in order to prevent accidental cross-fertilization. When flowers have bloomed and seeds are mature, cut off the heads and hang them in the bags to dry. Sow a few seeds from each plant. Seedlings from homozygous greens will be all green, those from heterozygotes will segregate green and albino. The remaining seeds can be used for class work or for breeding.

Tobacco (Nicotiana tabacum)   Segregation of seedlings will be seen about ten days after sowing if they are kept at 20–25 °C.

Blackened agar provides a good background for detecting the seedlings. Prepare as follows:

Add 30 g of agar and 6 teaspoonfuls of carbon black to 1000 cm^3 of water. Mix while boiling. Pour into glass Petri dishes and sterilize at 102 583 N m^{-2} (15 lbf in^{-2}) for 15 minutes.

The mixture can be sterilized separately and poured into pre-sterilized plastic Petri dishes. Allow the mixture to cool, but ensure that it does not solidify before pouring. Unsterilized but well washed glass or plastic Petri dishes can be used with little risk of contamination.

The albino seedlings do not survive. Future stocks can be obtained by the method described for barley. The plants are best kept in a greenhouse.

Tomato (Lycopersicum esculentum)   You might use the Packaged Genetic Lesson (PGL 4) on tomato cotyledon colour available from Practical Plant Genetics (see Chapter 2, page 38). The seeds will be sent with full details for culture in seed compost.

The tobacco and barley seeds are available from biological supply agencies.

5.3 The influence of chemicals on development

1 to establish that substances (that is, chemicals) in the cytoplasm can influence development

2 to show that some plant and animal hormones are examples of substances that affect the rate of development

Secondhand evidence is provided to introduce the hypothesis that the concentration of auxin determines the rate of development of a particular part of a plant.

An experimental procedure for testing this hypothesis is given in detail in the *Text*. It is intended that it should also form a basis on which the students can design their own experiments in section 5.4.

Apparatus and materials
Per pair:

Part 1 Instructions *1–3* in the *Text*
60–70 mustard or cress seeds
6 sterilized Petri dishes
6 discs of filter paper to fit inside Petri dishes
small tube of 1 per cent sodium hypochlorite
small bottle of distilled water

Part 2 Instructions *4–5* in the *Text*
6 Petri dishes each containing at least *6* cress seedlings with roots about 10 mm in length
tube of 10 p.p.m. auxin solution + sterile pipette (graduated 5 or 10 cm^3 type)
tube of 5 p.p.m. auxin solution + sterile pipette
tube of 1 p.p.m. auxin solution + sterile pipette
tube of 0.1 p.p.m. auxin solution + sterile pipette
tube of 0.05 p.p.m. auxin solution + sterile pipette
tube of distilled water + sterile pipette
sheet of millimetre graph paper

Indole–2–acetic acid (IAA) is used as the auxin (see Appendix 1, page 193).
If it is necessary to use a single pipette it should be well cleaned out with distilled water between each operation.

Part 3 Instruction *6* in the *Text*
6 Petri dishes containing cress seedlings in the auxin solutions or distilled water, set up in part 2
sheet of millimetre graph paper

(If necessary, materials can be shared between larger groups.)

Provided the students accept that the experiment meets the requirements of the hypothesis they should be able to perform it directly from the *Text*.

If a class timetable does not fit in with the timetable of experiments, *either* surface-sterilize and germinate the seeds for the students, *or* keep the seeds at a lower temperature in order to retard the rate of development.

The second part of the experiment, in which the seedlings are placed in auxin, can be extended for more than 48 hours, but careful watch should be kept on the material in case infection or other adverse effects set in. It is wise to germinate extra seedlings in a staggered series in order to overcome such snags.

Q1 Calculate the mean increase in length (separately for roots and stems) for each auxin concentration and the control

Q2 Express the effect of auxin on development by subtracting the mean increase in distilled water (control) from each of the mean increases in auxin. This tells us how much more (or less) the root or stem has developed *because of the effect of auxin.*

Q3 Have roots and stems shown a similar response to auxin concentration?

No, there are differences.

Maximum elongation of the stem occurs at between 0.1 p.p.m. and 1 p.p.m. Inhibition occurs at about 10 p.p.m. Maximum elongation and inhibition tend to occur at lower concentrations in the root (see also *figure 15*).

Q4 To what extent do the results support the hypothesis that the concentration of auxin determines the rate of development?

Here students should summarize their ideas:
a roots and stems are affected differently
b as concentration changes so does the effect on development (see *figure 15*).

Figure 15
A graph showing the usual type of relation obtained between auxin concentration and the extent of the stem's change in length.

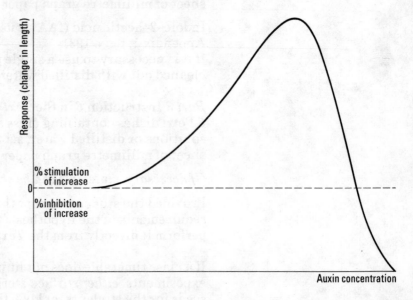

The perpetuation of life

Q5 Can you relate the results to
your knowledge of auxin and
tropic responses?

This question gives the teacher the opportunity to revise
tropic responses and to explain the paradox that auxin
tends to cause elongation in shoots and inhibition in roots
in relation to phototropism in shoots and geotropism in
roots.

Practical uses of plant hormones

The practical benefits derived already from work on plant
hormones are widely known and it is likely that even more
will come from future research. To follow up this topic,
use books and articles listed under 'Reference material'.

The topic lends itself to interesting project work. Students
could study the action of different proprietary plant
hormone preparations, by experiments. They could work
out the efficiency of different brands.

An attempt could be made to develop efficient selective
weed killers by studying the influence of different types of
plant hormones in varying strengths on different plants.
This latter project would give students an insight into the
way developments in applied biology take place, and it
would also emphasize the close link between 'pure' and
'applied' science. A possible scheme of work is as follows:

1 Determine some, or all, of the species of weeds and grasses
found on a lawn or field.
2 Determine the influence of different strengths of auxin,
for example, IAA, on the weeds and grasses grown
separately. From this work it is possible to deduce the
range of strengths which is most likely to give the best – not
necessarily ideal – results. It is likely that a compromise
may have to be reached. Thus, some grass species may
have to be destroyed in order to eradicate some weeds.
3 Select small sample plots on the lawn or field. Treat the
plots with different strengths of auxin. Now the influence
of natural conditions such as rainfall, soil drainage, type of
soil, light, etc., has to be considered. If at all possible the
influence of such variables should be allowed for in
determining the sample plots. Use some untreated plots as
controls.

Different methods of treatment, *e.g.*, spraying or pasting,
may also be tried out. When is the best time of day to treat
plants?

The results of such trials should indicate the most
efficient auxin strength to use. It may be feasible also to
test the influence of other hormones and chemicals.
A mixture might be found to be most suitable.

5.4 Animal hormones and development

1 to investigate, from photographs and a historical account, the effects of castration on vertebrates

2 to investigate metamorphosis and the importance of thyroxin, from first- or secondhand evidence

Castration and development

Students are asked to look at a series of photographs of Friesian cattle and then answer the questions.

Q1 What differences are there in *general body shape* between the cow and the bull (ignore sexual organs and mammary glands)?

The bull is much deeper in the chest region.

Q2 Which has changed most in *general body shape* during development from a calf: cow or bull?

The bull.

Q3 Does the steer look most like the calf, cow, or bull in general body shape?

The cow or calf. It does not have a deep chest like the bull.

Q4 Are your answers to questions 2 and 3 consistent with the hypothesis that the testes produce some hormone which affects the development of general body tissues?

Yes.

Students now read a letter by Gilbert White.

Q5 What is Mr Lisle's hypothesis?

That the removal of 'insignia' of masculinity (secondary sexual features) can have a castrating effect.

Q6 How would you test the hypothesis?

Pupils should discuss various ideas about removing secondary features and investigating the sexual behaviour and fertility of animals thus treated.

Q7 What is this evidence?

This section has already revealed evidence that the removal of 'insignia' of masculinity has little effect on later development or behaviour.

Gilbert White makes it clear that the horns of a bull are different from a cow's: a secondary sexual feature. But the account of development of Friesian cattle points out that calves are usually 'de-horned' and still develop into successful bulls or cows. (Men who shave, thus removing the secondary sexual feature of a beard, can still produce children!)

The perpetuation of life

Thyroid gland and metamorphosis

Students are asked to consider metamorphosis in the frog.

Development of limbs, loss of tail, etc.

Q8 What alterations to the body shape and structure occur as a result of metamorphosis?

Details of the practical investigation have not been included in the *Text*. Teachers may prefer to hold a discussion on experimental design and then present the class with secondhand data (as in *figure 16*) to analyse.

Figure 16
The effects, on the development of limbs, of administering different solutions of thyroxin, of iodine, and of thiourea, one solution each to different batches of ten *Xenopus* tadpoles. The treatment began at 7 weeks old and was continued for 10 days.

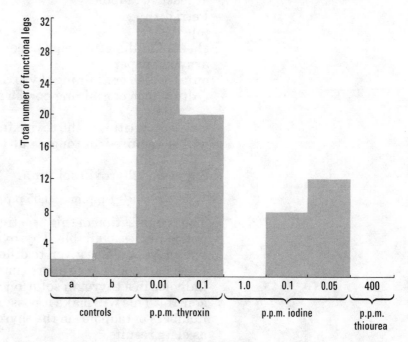

The experiment dealing with the influence of thyroxin on frog or toad metamorphosis is a clear demonstration of the role of animal hormones in development. It indicates that, up to a point, the amount of thyroxin determines the rate of metamorphosis. The experiment demonstrates only the influence of thyroxin additional to that which will be normally produced during development. Without exclusion of the thyroid, no proof can be offered which shows that thyroxin is essential for metamorphosis *per se*. A comparison of tadpoles with and without thyroid glands would be needed for this.

Tadpoles near to or at the beginning of the hind limb stage should be used for the experiment. Three to four weeks should be allowed for its completion. Frog tadpoles, if available, are suitable. Tadpoles of *Xenopus laevis* are also suitable and can be obtained from biological supply agencies throughout the year, or by injecting adults to induce them to lay eggs (Appendix 2, page 197).

It is as well to emphasize to the students at the outset that, because we are dealing with substances normally found in tadpoles, there is virtually no likelihood of causing them suffering provided living conditions are kept healthy and excessive doses are not used. Like plant hormones, animal hormones are often lethal when given in excess.

Apparatus and materials
Per pair:
at least *10* tadpoles
Petri dish
pipette
sheet of millimetre graph paper
drawing paper
culture dish containing thyroxin solution (1000 cm^3)
culture dish containing pond water (1000 cm^3)

Extra quantities of the thyroxin solutions and pond water will be required for topping up the dishes.

Suggested thyroxin solution:

        0.1 p.p.m.   0.05 p.p.m.   0.01 p.p.m.

The organization of this practical can vary with the time and facilities available. Thyroxin solutions of different strengths could be given to different pairs of students. Each pair could keep approximately two-thirds of its tadpoles in a thyroxin solution and one-third in the control. This will make it possible to compare similar numbers of tadpoles in the thyroxin solutions and control, as class results.

Of course, if there are facilities, a pair of students can study the effect of more than one strength of thyroxin solution. In this case the number of tadpoles given to each pair will have to be adjusted to ensure that over the whole class there are equal numbers of tadpoles in each thyroxin solution and in the control. There should also be enough tadpoles in each solution to make it possible to carry out tests of significance, if they are required.

Recording results

For observation, the tadpoles can be placed carefully in clean, small Petri dishes, without lids, containing just enough water to cover the animals when they lie flat on the bottom. A pipette can be used to keep the animals moist. The best way of lifting them from their culture dishes is to use the cupped palm of the hand or a small net. Do not use forceps or fingers as some tadpoles would inevitably get squashed that way.

Always impress upon students the importance of clean instruments, including hands, when handling animals.

Three types of observation can be made, using a hand lens.

1. Presence or absence of structure, for example, presence of limb buds, loss of horny jaws and operculum, appearance of ear drum (tympanic membrane). These changes need only be noted.
2. Changes of shape, such as enlargement of mouth, eyes becoming prominent, flattening of body dorso-ventrally. These changes can be recorded by drawing outlines of the body or parts of it, using the same scale in all drawings.
3. Quantitative changes, *e.g.*, in length of tail and in ratio of dorso-ventral and side-to-side dimensions of the body. Such changes can be measured by placing the Petri dishes on millimetre graph paper. The results can be plotted on a graph against time. The tadpoles can be steadied with the finger when measurements are being made.

Observations should be made as often as possible. They can be fitted in with other practical work. It is not necessary to obtain data from a large number of characteristics.

Measurement of the onset and rate of metamorphosis is affected by individual variation amongst the tadpoles. For this reason it is best to determine the percentage of tadpoles that have metamorphosed, at weekly or half-weekly intervals. A single criterion of metamorphosis, such as loss of three-quarters of the tail or presence of all four limbs, will have to be adopted in order that students can easily decide whether or not metamorphosis has occurred.

A test of significance could be applied to the difference in the percentage of animals metamorphosed in each of the thyroxin solutions compared with the control. To obtain adequate numbers for this calculation, use class results.

Inhibiting metamorphosis

Thiourea is known to inhibit the action of the thyroid gland. As a parallel experiment to the one using extra thyroxin, tadpoles could be kept in a solution of thiourea, and the postponement of metamorphosis in both untreated and thyroxin-treated tadpoles noted. Use a 1:2500 concentration of thiourea in pond water.

Thiourea should only be used in school with extreme caution and teachers should obtain permission from the school or the local education authority's Health Officer before carrying out this work.

A study of hormone influence on development poses the problem of determining experimentally that a particular substance exerts a control on the developmental processes, and it is inevitable that discussions of possible methods of doing this will arise during the work of this chapter. Jacobs (1959) has suggested certain formalized rules for tackling this problem and they are included here, with his permission, to afford a framework for such discussions.

His rules should not be accepted in a dogmatic way, but should be adapted to the level of attainment of the students and used to guide discussion.

'Types of evidence necessary to verify the hypothesis that "the development of structure S is normally controlled in organisms by chemical C".' Each type of evidence supports the hypothesis, and the more supporting evidence that can be found, the greater the likelihood that the hypothesis is correct.

'1   *Parallel variation*'
'Demonstrate that the chemical is normally present and that the amount of the structure varies in the intact organism in parallel fashion with the amount of the chemical.
If a quantitative relation can be demonstrated in such normal parallel variation, this strengthens the case.

'2   *Excision*'
'Remove the source of the chemical, and demonstrate subsequent absence or formation of the structure.

a   If a quantitative relation can be demonstrated between the amount of this artificial decrease of the chemical and the amount of decrease of the structure, this strengthens the case.

b   The chemical is sometimes removed by selecting genetic mutants.

c   A less direct – and therefore less satisfactory – method of meeting this requirement is to add other chemicals presumed, more or less specifically, to block chemical C, and to show that the structure is blocked too.

'3   *Substitution*'
'Substitute pure chemical C for the organ or tissue which had been shown to be the normal source of the chemical in the organism and demonstrate subsequent formation of the structure.

a   Quantitative evidence, as usual, strengthens the case. In the form of exact substitution, *i.e.* adding exactly the amount of chemical which is normally produced by the excised organ, it is particularly important when considering natural inhibition effects attributed to hormones. (Adding surplus hormones will unspecifically inhibit a variety of processes.) The ideal application of this rule would be to add a number of concentrations of the chemical – with one of the number providing exact substitution – and to demonstrate quantitatively parallel variation in the amount of the structure. Exact substitution should, of course, give exactly the normal amount of type of structure.

b Since the isolation of a naturally occurring chemical (particularly when it is a hormone) is a task which has often thwarted developmental physiologists, they may be driven to the following progressively less direct modification of the rule: substitute an extract of the organ; substitute chopped-up pieces of the organ; merely replace the excised organ with a different gap between it and the region where the structure is to form.

'4 *Isolation*'

'Isolate as much of the reacting system as possible and demonstrate that the chemical has the same effect as in the more intact organism. This reduces the probability that the chemical is acting primarily on some other process, or part of the organism, and only secondarily on the structure in question.

The part of biochemistry concerned with the *in vitro* reaction of extracted enzyme systems to added chemicals can be considered to represent the most extreme application of the isolation rule. Progressively less extreme isolation would be reactions of sub-cellular organelles; individual cells, tissues, pieces of organs, or whole organs (as in sterile cultures); leaf cuttings, etc.

'5 *Generality*'

'Demonstrate the generality of the results by showing that the other five points hold for species from a number of different families, as well as for the development of the structures in different kinds of organs.

'6 *Specificity*'

'Demonstrate that naturally occurring chemicals other than C have no such effect on the structure.

It is also important to remember that more than one substance may control a developmental process and, furthermore, that a physical process, such as the rate of diffusion, may be as important a controlling factor as a chemical which is involved with it.'

5.5 Measuring variation in development

Q1 Can you detect any differences between the pattern of development of different parts of the body?

Q2 Does the *rate* of development have an effect in different places?

Q3 Is the pattern of development of girls the same as that of boys?

Measurements of development illustrate clearly one of the characteristic features of living organisms – their range of variation. This is considered in more detail in section 5.5 of *Text* 1, *Introducing living things*.

The students' graphs of measurements of parts of the body against age should indicate the gradual increase in the relative proportions of the legs to the trunk as a person gets larger. The influence of puberty will be seen in the relatively greater increase in the size of the hips of the girls compared with boys after the age of 12. Comparison of the development of height and chest circumference indicates the overall greater rate of development amongst boys. The temporary increase in rate of development of the chest in girls between the ages of 11 and 14 is due, of course, to breast formation at puberty.
Explanations of the cause of different rates of development for different parts of the body can only be made tentatively. Hormones obviously have an influence. The effect of sex hormones should be clearly seen by the students.

The genetic influence on development is not clearly understood. There is enough evidence to make it reasonable to assume that endocrine effect is genetically controlled. It is also possible to consider that overall development likewise has a genetical component. The students will be able to accept that height, for instance, is probably determined to some extent by genes. They will probably have gained this knowledge from the reference to pea plants in Chapter 2. At this stage it is important to stress that in human beings the mechanism is much more complex.

The present-day view is that genes affecting the final size of an individual also secondarily affect the relative proportions of its parts. That these two aspects of development are related is indicated by the proportions of dwarfs and giants. Dwarfs tend to have the proportions of the early stages of normal development, *e.g.* large heads and small noses. Giants have the features you would expect of an extended normal development. Their heads are relatively small and their noses large.

So far in the course *mean* figures only have usually been used when comparing group results. This has allowed the results to be analysed without the need for additional mathematics that might have detracted from the main purpose of an exercise. At this stage the students should be sufficiently advanced to be able to consider the variation, or scatter of group results. The standard variation is an easily computed measure of variation.

When using the formula quoted in the *Text* the denominator $N-1$ gives a more accurate measure of variation than N, especially when small numbers are being considered. It is best to point that refinement out to the students when they understand the basic computations well.

Tests of significance

It may be thought appropriate to introduce the concept of statistical significance to students at this stage. Thus, when percentage or mean differences between groups are being analysed, it will be possible for students to distinguish between differences with a low probability of being due to chance error, which are thus more likely to be due to a specific factor, or factors, and those with a high probability of being due to chance error, which are thus less likely to be due to a specific factor or factors.

Students need not deal in detail with significance as a mathematical concept. This would detract far too much from the continuity of the course. Furthermore, it is probably more appropriate to mathematics lessons.

Probably the best idea to convey to the student is that tests of significance have been devised by which it is possible to distinguish whether differences in results are due to chance error, or to a specific factor or factors. Just as they have used staining methods without knowing the theory behind them, so, here, they are using a formula without knowing the theory behind it. In other words a statistical test is looked upon at this level in the same way as a food test. It is treated purely as a technique.

While it is not practicable to convey fully to students at this level the mathematical validation of statistics, it is important for them to realize that the practical aim of statistics is to reduce a mass of quantitative data to a few precise values that can be used for comparison and analysis. It is, in fact, a form of shorthand.

A second point that needs to be stressed is that statistical measures are of *samples* of population, not of whole populations. Thus, it is extremely important to assess the extent to which the sample portrays the whole population.

The means of different samples drawn from the same population are distributed about the mean of the population as a normal curve, which is the pattern exhibited by a distribution due to natural variation or chance. This distribution has its own mean and standard deviation. This standard deviation is the standard error (S.E.) of the population and can be estimated from the standard deviation of any sample by the formula:

$$\text{S.E.} = \frac{\sigma}{\sqrt{N}}$$

N = size of sample, σ = standard deviation of the sample.

The standard error indicates how much the mean of other similar samples drawn from the same population should vary. The essence of the tests of significance suggested for use in this course is that the standard errors of the two samples being compared are combined, and this standard error of difference is in turn compared with the observed difference between the percentage or mean score.

The standard error reflects the distribution you would expect if chance only were operative. Thus, the more similar the observed difference is to the standard error of the difference, the greater is the possibility that it is purely a result of chance. The more different these figures are, the less is the likelihood of its being due to chance.

Figure 17
A normal curve of population.

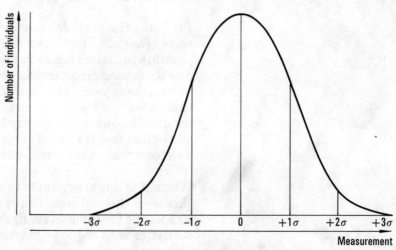

The perpetuation of life

Figure 17 shows a normal curve of the means of samples drawn from a population. Along its base is marked the position of the limits of curves drawn with the same population mean, but multiples of a standard deviation. It shows that at the extremes of the curve, where the standard deviation is high, few samples are included. Conversely, where the standard deviation is low, most of the samples are included.

The standard deviation for a population is its standard error. Thus, the smaller the standard error of a difference between two samples, the more likely it is that the difference is due to chance. A larger standard error gives less probability that the difference is due to chance.

In fact it can be calculated that the probability of an observed difference exceeding
S.E. × 1 by chance, is 1 in 3   S.E. × 3 by chance, is 1 in 370
S.E. × 2 by chance, is 1 in 22   S.E. × 9 by chance, is 1 in 17 000.

S.E. × 2, the accepted measure of significance, is located at the most sensitive point on the normal curve of a population, where steepness of slope changes rapidly.

The significance of differences between percentages

A possible approach, based on the experimental work in section 5.2, is as follows.

When you are counting large numbers it is usually convenient to turn them into percentages for comparison. Thus, suppose we were to sow a batch of tobacco seeds and keep them in the light, and another batch which we keep in the dark, we *might* get results like those in *table 4*.

	Total number of seeds sown	Number of albino seedlings	Number of green seedlings
Light	120	40 (33 %)	80 (67 %)
Dark	98	80 (82 %)	18 (18 %)

Table 4

You will see that it is much easier to compare percentages than actual numbers, especially when the totals are different. However, when making our comparisons, how can we be sure that any apparent similarities are not just due to chance? Do the differences we observe really mean something? In other words, are they *significant*? In our experiment with the tobacco seedlings, suppose 55 per cent green seedlings are produced in the light, and only 50 per cent in the dark. Are we right in concluding that there is a real difference between the two? Could it not be due to mere chance, as when we are dealt four aces in a hand of cards?

Various statistical tests have now been devised which enable us to calculate the *probability* that a given result is due to chance. This, then, leaves us with an arbitrary decision, as to when the influence of chance is so unlikely that it can be ruled out altogether for all practical purposes.

One of the ways of comparing two percentages is to calculate the standard error of the differences between them. This is an estimate of the probability that the difference is due to chance. To do this we apply the formula

$$\text{S.E.}_{\text{DIFF}} = \sqrt{\left(\frac{p_1 q_1}{N_1} + \frac{p_2 q_2}{N_2}\right)}$$

p_1 = percentage of one quantity of the first group
q_1 = the percentage $(100 - p_1)$ of the other quantity of the first group
N_1 = total *number* in the first group.

Thus, if we take the seeds grown in the light as our first group:
p_1 = percentage of albino seedlings = 33
q_1 = percentage of green seedlings = 67
N_1 = total number of seeds grown = 120

The second group consists of the seeds grown in the dark:
p_2 = percentage of albino seedlings = 82
q_2 = percentage of green seedlings = 18
N_2 = total number of seeds sown = 98

The standard error of the difference between the percentages =

$$\text{S.E.}_{\text{DIFF}} = \sqrt{\left(\frac{33 \times 67}{120} + \frac{82 \times 18}{98}\right)} = \pm 5.8$$

We are now in a position to compare our calculated difference between the percentages, as shown by the standard error, with the 'observed' difference obtained from our experimental results, that is, $82 - 67 = 15$ per cent.

The problem remains to decide whether a significant difference exists between them. The difference between the two quantities is generally judged to be significant when it equals or exceeds *twice* the standard error. This is the point at which the odds are roughly 20:1 against the results being due to chance. The reason for selecting $2 \times \text{S.E.}$ as the point where a difference becomes significant can be seen from the calculations we quoted on page 99.

To return to our example, the observed difference
$= 82 - 67 = 15$.

$$2 \times S.E. = 2 \times 5.8 = 11.6.$$

Since the observed difference is greater than twice the standard error, we can regard it as significant.

The significance of difference between means

The significance of the difference between the means of two different groups (1 and 2) can be measured by the standard error of the differences (S.E. $_{DIFF}$).

$$S.E._{DIFF} = \sqrt{\left(\frac{(\sigma 1)^2}{N_1} + \frac{(\sigma 2)^2}{N_2} \right)}$$

This formula loses accuracy if $N_1 + N_2$ is less than 30. Significance is assumed when the observed difference of the means equals, or exceeds, twice the standard error.

Whether or not tests of significance are incorporated into the work will depend, of course, on the mathematical background of the students, and the time available. It is surely important that students should be made aware of them to some extent.

Studying development in other organisms

The suggestions for further investigations at the end of section 5.5 are intended as leads into project work by interested groups or individuals. They are worth encouraging so that students can gain some ideas of the patterns of development in a variety of organisms. Information that should help with the projects can be found in other parts of this chapter (see the information on the influence of auxin on seedlings in section 5.3 and on the mass of babies at birth in the Background reading).

Suggestions for homework or preparation

The problems outlined in section 5.1 could be tackled as homework or preparation.

Other problems

1 Some leaves were kept in an atmosphere in which the carbon dioxide contained radioactive carbon. This carbon is used by the leaves for photosynthesis and to build up proteins and other substances just as the carbon in normal carbon dioxide is used.
One set of leaves treated in this way had a drop of auxin

placed on each leaf. Another set was untreated.
A photographic plate placed over a leaf will show up darker in areas where radioactive carbon has accumulated. When the two sets of leaves were treated in this way the untreated leaves produced a uniformly dark plate. In the leaves treated with a drop of auxin a dense, much darker patch appeared where the auxin had been placed. Otherwise these leaves were the same degree of darkness as the untreated leaves.
What conclusions can you draw from this experiment?

2 If the nucleus is removed from an *Acetabularia* plant and the 'hat' alone is cut off, a new 'hat' is formed. If, however, an *Acetabularia* is treated in the same way, but, in addition, the top of the 'stem' is removed, a new 'hat' is not regenerated.
Assuming there is a 'hat forming' substance in *Acetabularia* what can you conclude about its origin, what controls it, and where it is distributed?

3 In what sense can you say an organism is preformed? How does your view compare with the ideas outlined in the Background reading of Chapter 3?
(An organism can be considered to be preformed in the sense that it receives preformed genetic instructions from its parents.)

Q1 From the information in this account, complete *table 14* to show the development of the unborn baby.

Unofficial life

Age (months)	Length (cm)	Mass (kg)
0	0.01	—
1	0.70	—
2	1.3	0.001
3	7.0	0.030
4	19.0	0.170
5	—	—
6	35.0	0.550
7	—	1.000
8	—	1.800
9	50.0	3.300

Q2 Plot two graphs (on the same piece of graph paper, as indicated in *figure 104*) to show the pattern of development.

See *figure 18*.

Q3 How do the graphs compare with results for other work you may have done on development?

Students should relate the development of the human embryo to the development of barley seedlings (6.2) and children (5.3).

Q4 Which is the best parameter to use (length or mass) and why?

Mass gives a better idea of the increase in biological material (biomass) during the development of the human embryo.

Figure 18
Graphs of the development of the
human embryo.

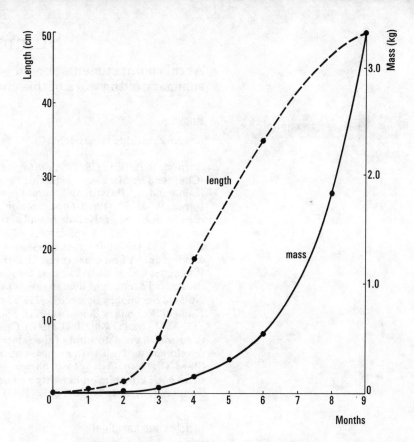

Q5 Collect data of mass at birth
among your brothers and sisters
or other relations and friends.
This information can be used to
test two hypotheses:
a That first babies are usually
smaller at birth than the later
children born to the same mother.
b That the mass at birth of a baby
is related to the size of the parents.

It will be important to collect data together from a number
of sources if proper assessment of the hypotheses is to be
possible. The hypotheses are based on 'commonly held
points of view'. Neither has much substance.

Q6 What other factors will you
need to consider in these surveys?

Factors such as: premature births, sex of baby, age of
mother, age of father. How is *size* of parents to be measured?
(It is probably best to measure height because mass of
parents is greatly affected by diet.)

Q7 Suggest other hypotheses that
might be interesting to investigate.

One hypothesis suggests that women who smoke during
pregnancy have smaller babies. Here one can remind
students of 'cause and effect'. If women who smoke have
smaller babies is it because smoking *causes* the baby to be
small? Or, is it because 'thin, nervous people' (who are
typically smaller) tend to have small babies and one of the
effects of being thin and nervous is that you tend to smoke.
The influence of the various factors suggested as answers
to question 6 would all make suitable ideas to investigate.

Summary

At this point students should be asked to make their own summary of the work of this chapter.

Books

*Reading suitable for students

Barnett, S. A. (Ed.) (1962) *A century of Darwin*. Heinemann. (Chapter 3 by Michie, D. 'The third stage in genetics', and Chapter 4 by Hammond, J. 'Darwin and animal breeding', are relevant.)
Baron, W. M. M. (1967) *Organization in plants*. Arnold. (Contains a concise account of the nature and action of plant hormones. The experiment with IAA and cress seedlings is based on one in this book.)
Ebert, J. D. (1970) *Interacting systems in development*. Holt, Rinehart & Winston. (A first-class revision of the subject.)
*Flanagan, G. L. (1963) *The first nine months of life*. Heinemann. (A clear, illustrated account of human pre-natal development.)
Nuffield Secondary Science (1971) *Theme 2 Continuity of life*. Longman.
Nuffield Secondary Science (1971) *Theme 3 Biology of man*. Longman.
Srb, A. M., and Owen, R. D. (1965) *General genetics*. W. H. Freeman. (Contains a good account of the relation between genetics and development; deals with nuclear control of development.)
Stern, C. (1960) *Principles of human genetics*. W. H. Freeman. (Contains three good chapters on heredity and environment.)
Tanner, J. M., and Taylor, C. R. (1967) *Growth*. Time–Life International.

Articles and pamphlets

*Reading suitable for students

Briggs, R., and King, T. J. (1957) 'Changes in the nuclei of differentiating endodermal cells as revealed by nuclear transplantation'. *Journal of morphology* **100**, 269–312.
*Jacobs, W. P. (1955) 'What makes leaves fall?' *Scientific American* Offprint No. 116. (Considers the role of plant hormones in leaf fall.
Jacobs, W. P. (1959) 'What substance normally controls a given biological process?' *Developmental biology* **1**, *6*, 527–54.
McMeekan, C. P. (1940) 'Growth and development in the pig, with special reference to carcass quality character'. *Journal of agricultural science* **30**, 276–343.
*Osborne, D. J. (1963) 'Hormonal control of plant death'. *Discovery* **24**, 31–5.
Parkes, A. S. (1947) '*Xenopus laevis*'. *School science review* **28**, *105*, 219–23.
*Salisbury, F. B. (1957) 'Plant growth substances'. *Scientific American* Offprint No. 110.
'Instructions for artificial induction of breeding in frogs', *and* 'Seedling mutants for the study of genetics'. (1963) Carolina Biological Supply Company, North Carolina, U.S.A.

Film loops

Biological Sciences Curriculum Study. 'Regeneration in *Acetabularia*'. John Murray.
Nuffield O-level Biology 'Removing and exchanging nuclei in *Amoeba*'. NBP–3. Longman.
Nuffield Secondary Science 'The growth of humans'. 0 582 24324 6 *and* 'Human hormones'. 0 582 24347 5. Longman.

6
Patterns and problems of development

Objectives of this chapter

1 to revise students' knowledge of different patterns of development (life-cycles)

2 to investigate the relationship between cell division and development

3 to show how different parameters of development produce different results

4 to re-introduce the problem, 'How do the cells of an organism develop into different forms even though, as a result of mitosis, they contain similar genetic instructions?'

5 to investigate the cells of a root tip in order to observe the effects of cell division, cell enlargement, and cell differentiation

6 to use work on root tips as an opportunity to *revise* mitosis

7 to consider the problem of how the cells of an organism are arranged together to produce multi-cellular organisms

8 to consider the control of development in whole organisms by a study of regeneration

This chapter brings some of the less understood and more controversial aspects of development to the notice of the students. It is hoped that this will encourage them to adopt a critical attitude during discussions on the subject. At the same time it will give an indication of the possible fields in which future discoveries may be made.

6.1 Different patterns of development

Objective

To revise students' knowledge of different patterns of development (life-cycles)

That different organisms display different patterns of development can be easily demonstrated with displays of life-cycles. It is essential that the students appreciate the variety of development before they study some of the more basic features and principles; otherwise they can gain a far too simple, and hence unrealistic, idea of the subject. It is very important for them to realize that their studies are introductory and by no means conclusive.

6.2 Is cell division related to the pattern of development?

Objectives

1 to investigate the relationship between cell division and development

2 to show how different parameters of development produce different results

A demonstration of different stages in the early development of a frog or toad up to the late blastula stage is probably the most suitable way of showing cell division.

A hand lens will usually provide the necessary magnification, but a binocular microscope would be preferable. For a class demonstration a micro-projector may be used. Living material, or material preserved in 70 per cent alcohol can be used.

By inducing animals to breed artificially (see Appendix 2, page 197), it is possible to demonstrate early development with living material. However, fitting this into the timetable is not always easy, and if it is attempted it is best done as a project running concurrently with the work of Chapters 5 and 6.

The soil nematode *Rhabditis* (collected as suggested in 2.1 of this *Guide*, page 23) may be used as a source of early stages of cleavage. Pick the nematodes off the dead earthworms or collect them by rinsing the dead worms in water in a cavity block or watch-glass. If some of them are mounted under a cover glass in a drop of water and gentle pressure applied with a needle to the cover, the females will liberate eggs in various stages from eggs to first stage larvae. The young worm consists of only about 200 cells – a useful point when discussing differentiation in 6.3. (See M. Wild's comment in *Nuffbiss,* No. 9.)

Figure 19
The curve of cell division.

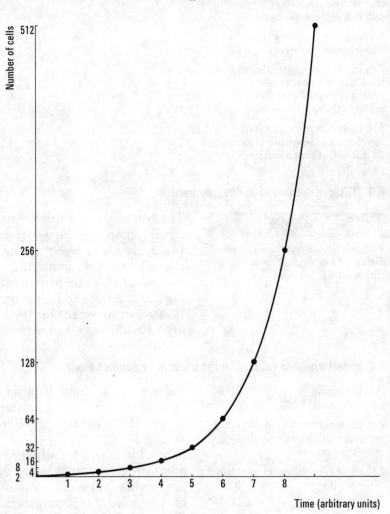

The perpetuation of life

Cell division and the measurements of barley plants

The graph of cell division in *figure 19* is an example of a theoretical model. The curve is characterized by increasing steepness as time increases.

For analysing the measurements of developing barley students need only consider the shape of the curves produced (*figure 20*). These can be used as the criteria of correlation. If curves are similar students can regard the phenomena they represent as correlated. Differences in the shapes of the curves are a measure of lack of correlation.

The curves for mass when wet and mass when dry show the greatest similarity to the curve for cell division. Mass when dry is probably the most accurate measure of development as it comes nearest to measuring the permanent mass of the plant.

Height is not a good measurement of development because one ignores the influence of girth, and using the number of leaves can also be criticized as merely representing a part of a plant. Nevertheless these are valid measurements of certain aspects of development. The curves for the development of both height and number of leaves are not similar to the curve of cell division.

Figure 20
Graphs showing the development of barley seedlings.

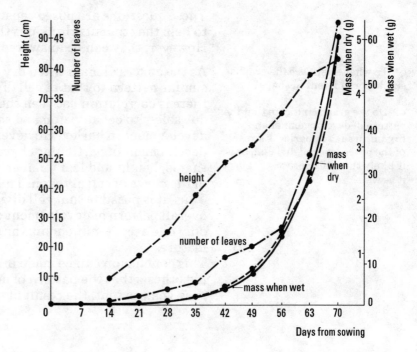

In analysing the results shown in table 15 in the *Text* the students should be made aware of some of the practical problems involved in such work which may well be possible sources of error. In removing the plants from soil, parts, particularly roots, may be broken off. Soil, if it is not thoroughly washed off, may distort measurements of mass. If too much time elapses between digging up and measuring, wilting may affect the results.

Finally, it should be noted that measurements were made on samples of plants at each time interval. The plants were then killed in order to obtain the mass when dry. Thus, the table does not take account of individual variation in the rate of development. This matter could be referred to in 5.5.

These practical problems need not detract from the general conclusions that can be drawn from the analysis. However, they can act as an incentive for students to try the investigation themselves as project work. Virtually any species of plant can be used and comparisons could be made between species. The work could be carried out in the school garden, in a greenhouse, or in a laboratory plant enclosure (see Appendix 2, page 206).

The graphs should be plotted as shown in *figure 20*.

These questions are posed separately in the *Text* in order to help the students think out their answers logically. However, they can be answered together.

Q1 To what extent are the graphs similar or different in shape?

Q2 Do the graphs indicate that the pattern of development of an organism such as barley is a result of the process of cell division, or is it likely that other processes affect it as well?

As the curves of mass when dry and mass when wet have a similar pattern to that of cell division, it would appear that there is correlation between the pattern of development due solely to cell division and the overall pattern of development in barley. However, the curves of the development of particular characteristics of the plants, that is, height and leaf number, do not follow the pattern of the curve of cell division. They show little correlation. Thus, it is possible that cell division may influence the overall pattern of development of the barley plant, but it is only one aspect influencing the development of certain features.
This conclusion can be made into the generalized hypothesis that the pattern of development of an organism is, to some extent, the result of the process of cell division.

Exponential series

It might be of interest to point out to the students that the series 1–2–4–8–16–32–64–128–256–n can be expressed as exponents to the base 2, as follows:

$$2^0-2^1-2^2-2^3-2^4-2^5-2^6-2^7-2^8-2^n.$$

To compute the number represented at any stage use the following procedure:

1. Find the logarithm of the base.
2. Multiply log base by the exponent.
3. Find the antilog of the result of *2*.

For example, to determine the number represented by 2^8

log of base (2) = 0.3010

log base × exponent (8) = 0.3010 × 8 = 2.4080

antilog 2.4080 = 255.9 = 256 (to nearest whole number).

6.3 How do cells differentiate?

Objectives

1 to re-introduce the problem, 'How do the cells of an organism develop into different forms even though, as a result of mitosis, they contain similar genetic instructions?'

2 to investigate the cells of a root tip in order to observe the effects of cell division, cell enlargement, and cell differentiation

3 to provide an opportunity to revise the essential features of mitosis

Examining root cells

Apparatus and materials
Per pair:
onion roots (see Chapter 3 and Appendix 2, page 219)
watch-glass
4 slides and coverslips
2 dissecting needles
1 scalpel or razor blade
2 forceps
filter paper (several sheets)
methylene blue stain
monocular microscope with $\frac{2}{3}$ and $\frac{1}{6}$ objectives

The roots should be stained with Feulgen (for details see section 3.3, page 44) immediately before the lesson. Alternatively, take fresh root material and macerate in M hydrochloric acid at 60 °C for 8–10 minutes during the first part of the lesson. The students then take the macerated roots and carry out the instructions in exactly the same way as if they had been stained with Feulgen.

If neither of these alternatives is possible, use previously fixed and stored onion roots (in acetic alcohol) and macerate in M hydrochloric acid at 60 °C for 10–12 minutes.

The practical work should be fairly straightforward. Students should be instructed carefully in the labelling of slides and the cutting up and identifying of the various pieces of root (figure 107*a* in the *Text*).

Q1 In which region, or regions, is cell division (mitosis) occurring?

In the one (or two) pieces nearest the tip.

Q2 Examine *figure 107b*, a diagram of a young root. When this root grows in length, are new cells added near the tip (region A) or near the base (region B)?

Near the tip, A.

Q3 For onion, $2n = 16$. What does this mean and how would the information help in your studies of cell division?

$2n = 16$ means that there are 16 chromosomes in normal onion cells. One might *hope* to count 16 chromosomes (each a double thread) at the beginning of mitosis and then count 16 single-strand chromosomes 'entering' each daughter cell.

Q4 Estimate any difference in the proportion of a cell occupied by the nucleus in the pieces taken near and far away from the tip.

The nucleus occupies a quarter to half the volume of the cell near the root tip. Further away the nucleus occupies a much smaller proportion of the cell: perhaps a tenth or twentieth.

Q5 Which of the following statements is the best summary of the changes you have seen?
As we move away from the tip of the root,
a the size of the nucleus increases;
b the size of the nucleus decreases;
c the size of the cell increases;
d the size of the cell decreases.

The answer is *c*: cell size increases.

The *Text* now offers the explanation that root development must involve cell division and cell enlargement.

Q6 If this is so we may expect the cells at the tip of the root to be smaller than those farther away. Is this the case?

Yes (as in question 5).

Q7 Are there other differences between cells found near to and farther away from the root tip?

The cells near to the tip are generally smaller and more uniform in size and shape. Their walls are thinner and their nuclei are frequently in mitotic stages.
By contrast, the cells farther away from the tip are larger, less uniform in size and shape, sometimes thick-walled, and unlikely to have nuclei in the process of division.

The completed table should be as follows:

Q8 Set out the results of your investigation in a table on the following lines . . . .

Stages of development	Description
stage 1: cell division	*mitosis at root tip*
stage 2: cell enlargement	*elongation in region about 2 mm behind tip*
stage 3: differentiation	*cells develop into a variety of different forms a few mm behind the root tip*

It is important for the students to realize that by cell division and enlargement the tip of a root is being continually extended. Thus, any one cell will, at successive time intervals, be placed further away from the tip (see *figure 21*).

The perpetuation of life

Figure 21
The relative position and form of a cell during the development of a root. (The cell is drawn much larger in scale than the outline of the root.)

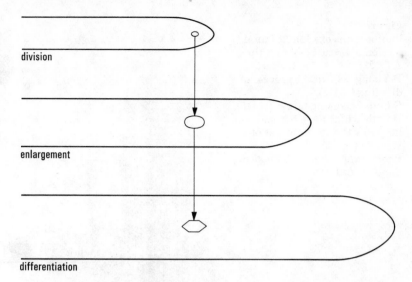

division

enlargement

differentiation

Thus, in its development a cell goes through three stages:
division – enlargement – differentiation

Figure 22 shows a number of sections of a root tip. It clearly shows the change in cell size behind the root tip. The figure can be used to confirm the class's results or to help those who missed or failed to complete the practical.

Marking root tips

Teachers may wish to demonstrate development in roots using the technique of marking the root tip. This could be set up in the same week as the microscope work of this section.

The stages of cell division and cell enlargement can be detected in a root by marking it at about 2 mm intervals for about 3 cm behind the tip. Leave the root to grow on the seedling or bulb.

After a few days, measure the distance between the marks. Those near the tip will not have changed, nor will those some distance behind the tip. The intervals just behind the tip will have increased. Thus, this is the zone of cell enlargement. The zone in front of it will be the zone of cell division and the one behind it the zone of differentiation.

Fingerprint ink or Indian ink is most suitable for marking a root tip. It can be applied by any suitable method, but care must be taken not to smudge the ink on the root. Only a very small amount need be used. As long as the marks are equidistant along the root when they are applied, the actual distance between them does not matter providing it is not more than 2 mm. (*Figure 23a* and *b*.)

Figure 22
Photographs of a longitudinal
section of broad bean root tip.
A (× 25).
B Young cells in the process of
dividing (× 115).
C Cells becoming larger (× 115).
D Cells which have become even
larger by the development of a
vacuole (× 115).
*Photographs, Harris Biological
Supplies Ltd.*

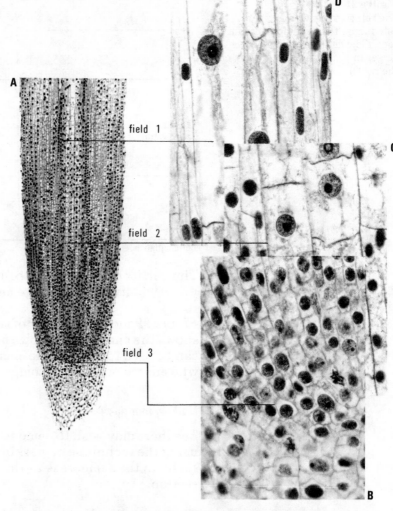

field 1

field 2

field 3

It is not satisfactory to use a single strand of cotton or a
pen for marking, because it is difficult to gauge the distance
between marks. An old comb can be used, or strands of
cotton wound round the threads of two bolts and stretched
taut. See *figure 23c*.

Figure 23
Determining the zones of
development in a root tip by
marking.
a Original markings.
b Marking after several days'
development.

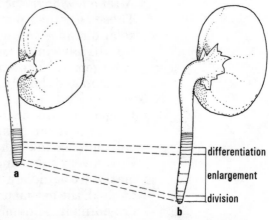

differentiation

enlargement

division

a

b

Figure 23c
One type of root tip marker.

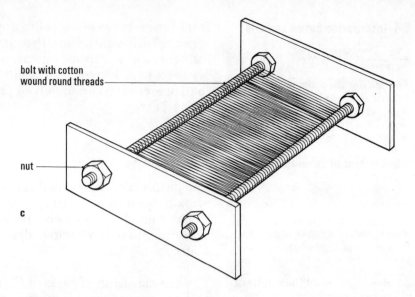

bolt with cotton
wound round threads

nut

c

This section ends with a restatement of the unsolved problem. We have *described* cell development in terms of three stages but we have not been able to *explain* how differentiation occurs in genetically identical cells.

Presumably differentiation is the result of a process, or processes, superimposed on the genetic mechanism for protein synthesis.

It would appear that at some stage, or stages, in this basic mechanism, the chemical and physical environment of the cell must be influential. Presumably a 'feedback' mechanism must operate because the cell environment itself must be determined to a large extent by the genes.

Cytoplasmic influence on the nucleus has been demonstrated and the students will have studied this in Chapter 5. It is possible that the environment of differentiating cells influences the pattern of differentiation. Embryonic cells show this when they are cultured *in vitro*, for, sometimes, they do not continue to differentiate. Various chemicals have been shown to influence the pattern of differentiation shown by such embryonic cells.

The idea that the interaction of genes *may* influence differentiation will seem credible to the students, for they have already seen how the interaction of genes can influence the production of a characteristic during their studies in Chapter 2 when they dealt with dominance. Nevertheless, it should be remembered that these types of interactions do not result in differentiation.

6.4 Interaction between cells

Objective

To consider the problem of how the cells of an organism are arranged together to produce multi-cellular organisms

Interaction between cells during development is another aspect of differentiation. Presumably it is one of the main influences at the tissue and organ levels of organization. However, it is not possible to do much more than present it as an unsolved problem in the way outlined in the *Text*.

6.5 Control of development in the whole organism

Objective

To consider the control of development in whole organisms by a study of regeneration

Q1 Bearing in mind the results of these two experiments, can you suggest a hypothesis to explain how the anterior–posterior arrangement of an organism is determined?

The problem of the overall control of development has already been considered to some extent in Chapters 5 and 6. Genetic and hormonal control have been mentioned. The experiments with *Planaria* described in the *Text* introduce a further aspect.

The regeneration of parts of *Planaria* indicates that the ability to regenerate is distributed in an anterior–posterior gradient. The relative rates of disintegration of the tissues of *Planaria* by metabolic poisons indicate that the metabolic rates of the tissues are arranged along a similar anterior–posterior gradient.

This correlation would indicate that there may be a causal relation between the two. It suggests the hypothesis that the form of the animal is determined, at least in part, by a pattern of different rates of metabolic activities in its body.

This hypothesis in its turn implies the question, 'What determines the pattern of the different metabolic rates?'. Unfortunately, no answer is available. There may well be a genetic control but it is likely to be much more complex.

A discussion of these experiments with *Planaria* could lead to a wider consideration of the subject of axial fields. For this, more advanced texts should be consulted.

Regeneration experiments with *Planaria*

Regeneration experiments with *Planaria* are not always easy. It must be expected that a proportion of the animals will not survive the operation. Extreme care must be taken to keep materials clean.

However, it is an exercise worth attempting, for one successful case of regeneration will arouse considerable interest among students.

A thoroughly clean, stainless steel, razor blade can be used for cutting the animals. Part of the cutting edge is broken off and inserted into a short wooden handle. This improvised scalpel makes an easily manipulated instrument.

The operation can be performed on a glass slide. The animal should be kept well moistened with water. A hand lens or microscope is needed.

After the operation the parts of the organisms are kept separately in regularly cleared water without food. Full regeneration usually takes up to three weeks.

The most important thing is to change the pond water the day after the operation.

The experiment described in the *Text* could be performed partly by the class, partly as a demonstration. In order to show the effect of a metabolic poison, an animal should be placed in 0.001 M potassium cyanide solution. As, of course, metabolic poisons are very dangerous, it is probably wisest not to use them for class work at this level. Thus, this part of the experiment should be demonstrated by the teacher.

Further experiments prompted by the questions in the *Text* could be attempted as project work by students.

Q2 What will happen if part of one planarian is grafted on to another?

Transplant tissue from the anterior region of one animal to the posterior region of another. This can be done by removing a circle of tissue from the posterior region of an animal. If a piece of tissue from the other animal is placed in the space of the circle it should then graft on.
Success in this experiment is, however, limited and several attempts will probably have to be made. Use animals of the same species for grafting.
A second head will develop from the grafted piece, showing the dominance of anterior tissue over posterior tissue.

Q3 Is there likely to be a gradient across a planarian from left to right? What would you expect to happen if a planarian were cut lengthways?

Remove the head and make a lengthwise slit through part of the body. Two heads will develop because of the equal dominance of the split ends.

Q4 Does the size of the part of a planarian which is cut off have any influence on its powers of regeneration?

Cut a very narrow slice of tissue from the anterior end of a planarian. Two heads develop. Apparently a narrow slice of tissue has no anterior–posterior gradient and cannot form a tail.

For notes on the culture of *Planaria*, see Appendix 2, page 218.

1 to show that the ability to regenerate a whole organism from a small portion of a differentiated adult can occur in the plant kingdom

2 to suggest a practical investigation which might be an alternative to work with *Planaria*

3 to provide a link with the work of Chapter 7, 'Different ways of breeding', where vegetative propagation is described

Q5 How can the cells of a leaf 'know' how to make a whole *Begonia* plant?

Q6 How can the cells of a *leaf* differentiate into roots and stem?

Q7 How would you test the following hypotheses?
a Begonia leaf cuttings will only develop if the cutting includes a piece of vein.

b Begonia leaf cuttings show a *polarity* and new plants will only develop if the leaf cutting is planted the 'right way up' (that is, proximal or basal region in the soil and distal or tip end of leaf out of the soil).

Problem

Pupils are told that some varieties of *Begonia* plants (for example, *B. rex*) will develop from a small cutting, about the size of a postage stamp, taken from a leaf.

The cells of the leaf are derived by mitosis from the zygote. Therefore, they may be expected to carry identical instructions to those found in the original zygote cell from which the whole plant developed.

This question raises the fascinating problem of 'de-differentiation'. The various cell types in the leaf may perhaps only develop into similar cell types in stem or root. Alternatively, a few cell types in the leaf may have the ability to change their structure and function.

Presumably any piece of leaf the size of a postage stamp will have a number of small veins. More probably, there must be a *major* vein rib running the length of the cutting (perhaps to distribute food and water). A suitable experiment may be to take about ten leaf cuttings with a major vein and ten without. Plant these in compost at suitable temperature (about 20 °C) and fairly humid. The leaf cuttings should *all* be planted with the correct 'polarity' (see hypothesis *b*).

A suitable experiment would be to take 30 leaf cuttings, each with a major vein. Plant ten the right way up, ten upside down, and ten on their side.

These investigations with *Begonia* leaf cuttings would make excellent projects, long-term investigations, or demonstrations by the teacher. Difficulties may be encountered in propagation. Conditions need to be warm and humid: this unfortunately, may favour various fungi. Use sandy soil or vermiculite as a growth medium.

Suggestions for homework or preparation

Much of the work in sections 6.2 and 6.5 could well be set for homework or preparation.

Q1 Imagine you are collecting money to be used for cancer research. A wealthy business man, who might be expected to make a large donation, raises the following objection: 'I don't see much point in my giving money to your charity; I have heard that it is spent on studying carrots growing in the laboratory!' Explain to this man why his gift, even if used to finance studies of carrot development, can be helpful in cancer research. Use only simple scientific language and present your case in not more than 300 words.

Carrots, coconuts, and cancer

The short essay is intended to give students an opportunity to express their ideas about the relevance of pure research in relation to applied problems. Their knowledge, at this stage, is limited, but the Background reading is a clear account of the link between 'unlikely' topics: carrots, coconuts, and cancer.

Summary

At this point, the students should be asked to make their own summary of this chapter.

Books

*Reading suitable for students

*Brenner, S., Curtis, A., Smith, J. M., and Wolpert, L. (1963) *Cells and embryos*. BBC. (Up-to-date and contains some interesting speculations.)
Carter, G. S. (1951) *A general zoology of the invertebrates*. Sidgwick & Jackson. (Contains a very good survey of some of the problems of development. A good section on regeneration.)
Ebert, J. D. (1970). *Interacting systems in development*. Holt, Rinehart & Winston.
Gliddon, R. (1970) *Units of life: cells and unicellular organisms*. Heinemann Educational.
Hamburger, V. (1960) *A manual of experimental embryology*. University of Chicago Press. (Contains details of a wide range of investigations in experimental embryology. A first-class reference book.)
Mellanby, H. (1963) *Animal life in fresh water*. Methuen. (For identification of planarians.)
Scientific American Single-topic Issue (1961) *The living cell*. W. H. Freeman. (This collection of 9 articles includes those by Fischberg and Blackler, Mazia, and Moscona, which are listed as separate offprints below. It is worth getting the complete collection.)

Articles

*The following *Scientific American* Offprints are recommended:
Fischberg, M. and Blackler, A. W. (1961) 'How cells specialize'. No. 94.
Mazia, D. (1961) 'How cells divide'. No. 93.
Moscona, A. A. (1961) 'How cells associate'. No. 95.
Singer, M. (1958) 'The regeneration of body parts'. No. 105.
Steward, F. C. (1963) 'The control of growth in plant cells'. No. 167.
(A fuller account of Steward's investigations: see Background reading.)
Waddington, C. H. (1953) 'How do cells differentiate?' Offprint No. 45.

Outline of the study of breeding
(Chapter 7)

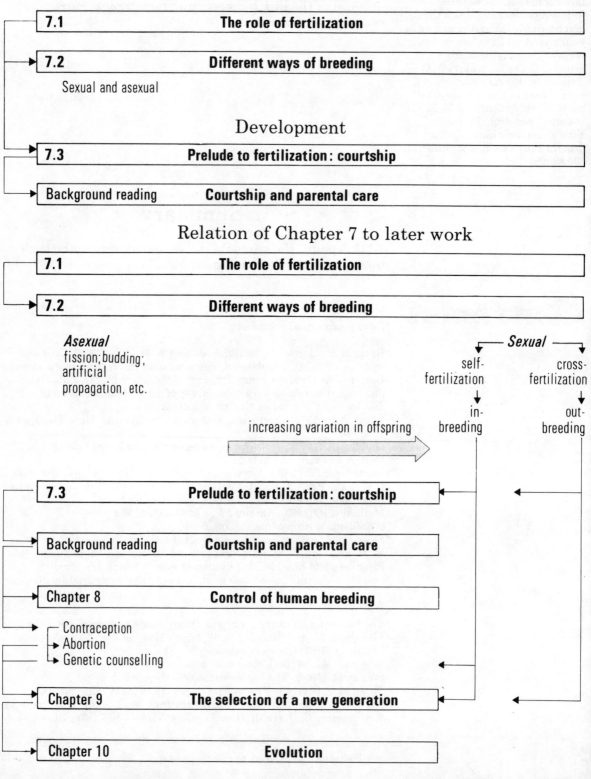

Introduction

7.1	**The role of fertilization**

7.2	**Different ways of breeding**

Sexual and asexual

Development

7.3	**Prelude to fertilization : courtship**

Background reading	**Courtship and parental care**

Relation of Chapter 7 to later work

7.1	**The role of fertilization**

7.2	**Different ways of breeding**

Asexual
fission; budding;
artificial
propagation, etc.

Sexual

self-fertilization   cross-fertilization

in-breeding   out-breeding

increasing variation in offspring

7.3	**Prelude to fertilization : courtship**

Background reading	**Courtship and parental care**

Chapter 8	**Control of human breeding**

- Contraception
- Abortion
- Genetic counselling

Chapter 9	**The selection of a new generation**

Chapter 10	**Evolution**

Different ways of breeding

Objectives of this chapter

1 to illustrate the distinctions between asexual and sexual reproduction, self-fertilization and cross-fertilization, in-breeding and out-breeding

2 to illustrate the advantages and disadvantages of the above forms of reproduction to organisms living under natural conditions, and their application to animal and plant breeding by man

3 to investigate courtship behaviour in various animals including humans

4 to develop an understanding of different aspects of parental care: physical, educative, and emotional

Introduction
(Section 7.1)

7.1 The role of fertilization

This section emphasizes the part played by fertilization in the production of variety among living organisms. The distinction between sexual and asexual reproduction is also drawn and this provides a lead into the next section which aims to show that the methods of reproduction (that is, breeding) employed by animals and plants affect the variety to be found among their offspring.

7.2 Different ways of breeding

Objectives

1 to help students to understand the different nature of reproductive methods

2 to emphasize the importance of different methods of reproduction to living things

Q1 What is the difference in the amount of variation shown by the individuals of the two populations?

Q2 Do the results shown in the graphs support the idea that sexual reproduction increases variation in a species?

1 Self-fertilization and cross-fertilization

Asparagus
Figures 115 and 116 in the *Text* describe asparagus shoots grown from seeds produced by cross-fertilization and by self-fertilization.

The population bred by cross-fertilization shows more variation than the population bred by self-fertilization.

Sexual reproduction, as a result of cross-fertilization, allows the re-combination of genes from two individuals to give a wide variety of shoot lengths. Sexual reproduction, as a result of self-fertilization, allows the re-combination of genes from only one individual which, therefore, results in a smaller variety of shoot lengths.
Sexual reproduction does increase the variety of life. Students may also mention that gamete formation allows

for some rearrangement of genetic material resulting in variation. Mutation of genes, although very unlikely, also gives rise to variation.

Primroses

Q3 How does pollination take place between the pin-eyed and thrum-eyed forms?

Cross-pollination leading to cross-fertilization is favoured between the pin-eyed and thrum-eyed forms. The insect must go from pin to thrum or thrum to pin to achieve cross-pollination. Students should make some mention of the significance of the position of the stigma and anthers in the two forms and observe that an insect with a long proboscis must visit the flower to effect pollination. There is also a difference in the size of pollen grains between the two forms which helps to promote cross-pollination.

Q4 How does pollination take place in the homostyle form?

Self-pollination leading to self-fertilization is favoured in the homostyle form. When the insect visits the flower, pollen will be transferred from the stamens to the stigma as the proboscis is inserted into the corolla tube in search of nectar. Maps A and B in figure 118 in the *Text* are hypothetical, but are based on research which has shown the spread of the homostyle form of the primrose.

Q5 Why has the distribution of the homostyle form increased?

Cross-pollination will not be achieved every time in the pin- and thrum-eyed forms; the insect may go from pin to pin or thrum to thrum. Self-pollination should be more successful in the homostyle form, resulting in fertilization and a higher proportion of seeds produced than in the pin- or thrum-eyed forms. This will result in a greater dispersal of the homostyle form.

2 Dog breeding

Figure 119 in the *Text* shows a cross between a bulldog and a basset hound.

Q6 Can you explain the results of the cross?

Re-assortment of genes will take place during gamete formation and fertilization, giving rise to dogs in the first generation with characteristics of both parents, bulldog and basset hound. Because of the dominance of some genes, certain characteristics of the parents may not appear in the first generation. A further assortment of genes as a result of crossing two members of the first generation gives a wider variety of forms in the second generation.

Q7 How would you set about breeding a number of dogs similar to one of the types produced in the second generation of the cross?

Breed further numbers of the second generation until several animals near to or similar to the desired type are produced. Then breed only from these animals, continually selecting only those forms that are similar to the desired type.

The perpetuation of life

This problem should bring out the differences between in-breeding, in which offspring are produced from closely related parents (bulldog × bulldog), and out-breeding, in which the offspring are produced from unrelated individuals (bulldog × basset hound). It also emphasizes that sexual reproduction by cross-fertilization affects the amount of variation produced. The idea of selection could also be introduced at this point and the use of in-breeding coupled with selection to produce a desired type.

3 Asexual reproduction

The *Text* gives a selection of examples to illustrate asexual reproduction. Most of the examples are easily available and can be demonstrated as living or preserved specimens. Some short- and long-term experiments could be set up to amplify the *Text*.

Short term
The growth of bacterial colonies on agar plates would illustrate the rate of asexual reproduction by simple fission. Some discussion of the factors that stop the unlimited multiplication of the bacteria would be relevant; lack of food, build-up of waste, and overcrowding.

Two yeast suspensions, one with sugar and one without sugar, can be set up 24 hours before the lesson. On examination under the microscope only the yeast suspension with the sugar should show budding yeast cells.

Pleurococcus from the trunk of a tree may also show cell splitting under the microscope. The sporangia of *Mucor* and the conidiophores of *Aspergillus* and *Penicillium* can also be easily demonstrated under the microscope.

Long term
Bryophyllum plants are easily grown from the leaf plantlets. Strawberry plants can be grown from runners. Allow the end of the runner to root in a small plant pot buried in the ground near the parent plant. When the daughter plant is well rooted the runner can be cut and the pot removed from the soil.

Sections of couch grass rhizome can be planted in pots and new shoots will develop. This shooting will also occur with sections of dandelion root.

4 Artificial propagation

Students can do simple cutting experiments with geraniums; the technique is shown in *figure 24*. The use of a hormone rooting powder is recommended as it does increase the percentage of those that root. A sandy potting soil should always be used for cuttings. Willow twigs placed in a bottle full of water will develop roots and can then be planted.

The leaves of *Begonia* can be placed on potting soil after the midrib has been cut across in several places and dusted with hormone rooting powder (see also page 116).

(see also page 116)

Figure 24
Taking a cutting from a geranium plant. With a sharp knife cut the shoot across at A–B. Remove the bottom leaves C and D of the shoot.

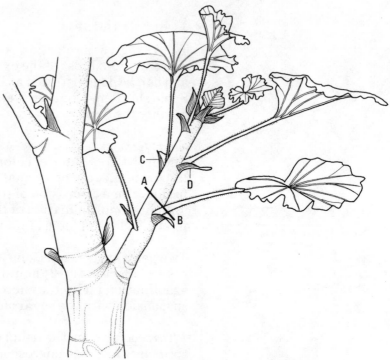

One reason why rose growers do not grow roses from seeds is that germination is difficult as the seeds require a dormancy period with temperatures near freezing.

The technique of budding roses is illustrated in *figure 25*, and can be quite successful with practice. Some budded roses in the school garden would be useful demonstrations. The season for budding is from June to early August.

q8 Why do you think gardeners choose these methods of reproducing their plants rather than growing them from seeds?

Asexual methods, such as cutting and grafting, limit the amount of variation produced: new combinations of genes are not possible as there is no gamete formation or fertilization. In this way a gardener can maintain a uniform strain of plant and there will also be less wastage than would be expected with sexual reproduction.

Figure 25 (right)
The technique for budding roses.
After R. Stone.

The perpetuation of life

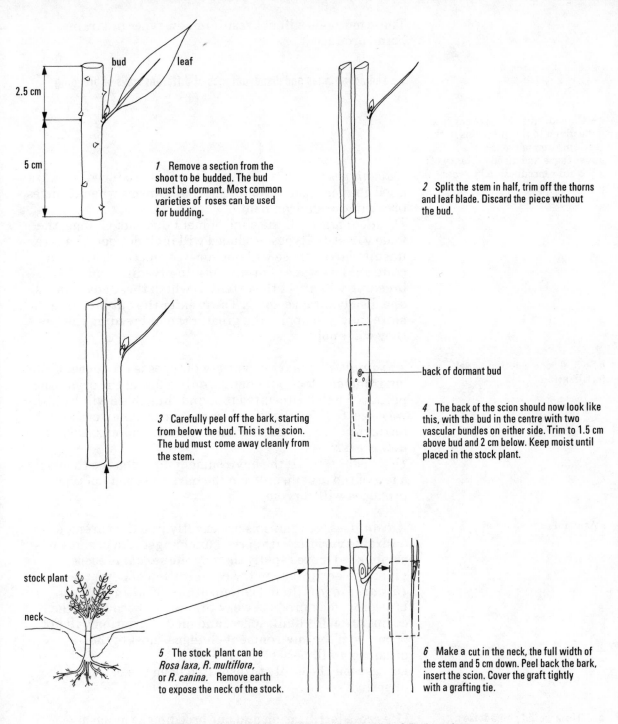

bud   leaf

2.5 cm

5 cm

1   Remove a section from the shoot to be budded. The bud must be dormant. Most common varieties of roses can be used for budding.

2   Split the stem in half, trim off the thorns and leaf blade. Discard the piece without the bud.

3   Carefully peel off the bark, starting from below the bud. This is the scion. The bud must come away cleanly from the stem.

back of dormant bud

4   The back of the scion should now look like this, with the bud in the centre with two vascular bundles on either side. Trim to 1.5 cm above bud and 2 cm below. Keep moist until placed in the stock plant.

stock plant

neck

5   The stock plant can be *Rosa laxa, R. multiflora,* or *R. canina.* Remove earth to expose the neck of the stock.

6   Make a cut in the neck, the full width of the stem and 5 cm down. Peel back the bark, insert the scion. Cover the graft tightly with a grafting tie.

7   Replace the soil over the graft and water. Leave until the following spring and then cut off all top growth as close to the bud as possible, leaving only the bud and the neck. This bud will develop into the new rose bush.

These methods will not result in new types or strains being produced.

5   The advantages and disadvantages of different ways of breeding

Q9 Considering the answers you have given to the questions in this section, summarize the advantages and disadvantages of:
a Sexual reproduction by cross-fertilization

Advantages   A wide variety of types is produced, giving a good chance of survival for some organisms when changes occur in the environment.
Disadvantages   If the environment does not change, the wide variety of types produced will include some that are unsuited to it: these will not survive and thus there is a continual wastage of organisms. In-breeding and out-breeding will affect the extent to which these advantages and disadvantages apply. The greater the out-breeding the more they will apply; the greater the in-breeding the less they will apply.

b Sexual reproduction by self-fertilization

Advantages   A narrow variety of types is produced. If the environment does not change, only a few of the organisms produced will be unsuited to it and thus there will be little wastage. Self-fertilization is more certain than cross-fertilization, giving an advantage in the colonization of new areas.
Disadvantages   If the environment does change then only a few of the organisms from the narrow variety of types produced will survive.

c Asexual reproduction.

Advantages   Organisms are exactly like the parent, so desirable varieties are spread unchanged. An organism is established more rapidly than by the sexual process, giving an advantage in the colonization of new areas.
Disadvantages   Rapid colonization, resulting in over-crowding, may produce weak organisms because of the competition for light, space, and food. Variation will not occur, so if the environment changes, most of the organisms will perish.
Disease may be easily transmitted from the parent to the offspring.

Q10 How would you advise a nurseryman or gardener to use these different ways of breeding?

Use cross-fertilization and out-breeding as much as possible to produce new varieties. When a desired type is obtained, use self-fertilization, asexual reproduction, and artificial propagation to maintain the new type in successive generations.

Development
(Section 7.3 to end of chapter)

7.3 Prelude to fertilization: courtship

Objectives

1 to indicate the variety of stimuli which can result in copulatory behaviour and courtship

2 to attempt to distinguish any survival value in courtship behaviour

3 to emphasize the intellectual pitfalls in drawing parallels between human and animal behaviour in courtship

4 to establish that courtship behaviour can become ritual

This is a very complex subject to introduce at this level and some knowledge of animal behaviour is necessarily assumed (*Text* 3, *Living things and their environment*, Chapter 12).

The use of secondhand evidence does not necessarily mean that such evidence is second-rate. For studying courtship, the accurately observed films suggested for use are the best and most convenient form of evidence available, but can never fully describe all the variables involved. While, possibly, they may not encourage primary discovery, they lend themselves particularly to the objectives of critical interpretation and formation of hypotheses.

1 Courtship in spiders

The film recommended, 'Comparative courtship' (see the Reference material on page 129), is one of a series on the biology of spiders. This series includes two other films, 'Comparative courtship in Lycosidae (wolf spiders)' and 'Comparative courtship in Salticidae (jumpers)', which teachers may like to use, either in preference to the recommended one or as an extension of the work. (Again, see the Reference material.)

Q1 How do the sexes actually come together? What stimuli make one spider leave its web and seek another?

Q2 What problems are involved in the mating procedure – what behaviour patterns must not be released if fertilization is to take place? How do the different spiders solve this problem?

Q3 What are the limits of accuracy of this filmed evidence?

Full answers to the three questions given in this section depend on which films are used, and should be self-evident.

2 The three-spined stickleback

The filmed evidence, 'Behaviour of the three-spined stickleback' (see the Reference material, page 129), is not only descriptive but also experimental. It may be needed only to reinforce work on animal behaviour done in *Text* 3, *Living things and their environment*, Chapters 11 and 12, using, possibly, a different frame of reference. A highly coloured diagram could be used instead of the film if this work has already been covered thoroughly. The diagrams in Tinbergen, 1966 (see the Reference material on page 129), are suitable.

Although answers to questions on animal behaviour can rarely be precise, and must always be hedged with 'as far as we know', some points are given below which should at least be expected in answers to questions on the film of stickleback behaviour. Questions should be asked to elicit the gaps in knowledge as well as those to which a ready answer can be given.

Q4 What kind of stimuli could cause the males to leave the shoal?

This is an unexplored region at the time of writing. Possible hypotheses include:
a Some internal change takes place which stops the male fish from responding to the stimuli which normally keep him in the shoal.
b There are repellent stimuli which lead the shoal to disperse.
c There are internal changes of perception so that stimuli offered by appropriate nest-building sites are selected rather than those offered as an inducement to keeping in the shoal. Since hormonal changes must be present to produce the change to breeding colours, imminent when the fish leave the shoal, it is possible that they also provide the mechanism which initiates solitary behaviour. On the other hand, all the stimuli needed could be provided by the pregnant females – the hypotheses are endless, and in a class of a high enough level of ability, this problem can prove intellectually exciting.

Q5 Why should spring mating be so prevalent among many animals?

The birth of young often coincides with the larger quantities of vegetable food available in spring, as the amount of light energy increases. The time of mating is related to gestation periods and those animals with shorter gestation periods actually mate during the spring itself.

Q6 What is the advantage of territorial boundaries for each nest?

Probably as an advance defence preparation.

Q7 What is the precise stimulus which causes the male to drive other males from his nest territory?

The colour red.

Q8 What are the precise stimuli which release *a* the courtship dance and *b* mating behaviour?

a A pear-shaped model of the same size as a female stickleback – known as the 'sex bomb' by the research team.
b Any tail-shaped piece will induce the male to press his snout against it, as he does to promote egg-laying in the female. When the male enters the nest after the female has

left, he fertilizes the eggs – but the precise stimulus which makes him do so is not known.

Courtship and parental care

The Background reading considers the nature of courtship in certain animals, including man, and its possible value for survival. Its other object is to consolidate earlier work on parental care. Where this was mainly concerned with the physical aspects, students are now asked to consider the social and psychological implications.

Courtship in birds

There are a large number of films available on courting behaviour of birds, and teachers who wish to emphasize this part of the work can select those suitable for their own interests. None have been suggested here as the subject is very complex and hence only a brief treatment, suitable for this level, is outlined.

Courtship in man

Since courtship in man is infinitely more complex than in birds, it may seem illogical to include it at this level at all. Also, projects in other disciplines may be felt to be more suitable for studying this phenomenon, such as the two packs, *Relationships between the sexes* in the Schools Council Humanities Project and the *Lifeline* series in the Schools Council Moral Education Project. (See the Reference material on page 129.)

Nevertheless, adolescents are very curious about themselves and some attempt to clear away the confusion surrounding the biological basis of human courtship behaviour may be felt to be worth while. The work needs to be based on a comprehension of the simpler forms of courtship behaviour outlined already. It might take the form of a discussion of filmed evidence and personal observation. It may provide an opportunity for team teaching with humanities or social science departments; it is often felt that studies of man benefit from the presence of a biologist!

Useful teaching aids, of which details are given in the Reference material on page 129, are 'Married life' and 'Teenagers' from the series of films, 'The family of man'. The film loop, '*The engagement*', is very artificial, but this is one of its strengths, as it provokes laughter and ensures a good communicating atmosphere for its discussion. It shows a rather trite situation, but this does away with the need for anecdotal embellishment and makes it easier to see clearly the points under examination.

Teachers may wish to use questions on work sheets for the interpretation of the evidence of the films suggested in this section, to be discussed in small groups with the class coming together before the end of the lesson for some form of consensus record to be made. If discussion with a whole class is attempted, many students will not find the opportunity to express and refine their opinions.

Parental care

It is assumed that simple ideas about parental care will have been comprehended in earlier work, in particular, the work outlined in *Text* 1, *Introducing living things*. This treatment, therefore, seeks to consolidate earlier work and to extend it to consider the social and emotional aspects of parental care in man. The analysis of parental care in man is a very inexact science and no attempt has been made to cover all its aspects. It is included here merely as an aid to students' observations of the behaviour of their own species.

Physical parental care

Two loop films are suggested to consolidate earlier work. They are 'Parental care in wild animals' 1 and 2, (see the Reference material on page 129). Suggestions for work sheets are included with each loop, and teachers may wish to adapt these for work with small groups as an alternative to teaching the material with the entire class. When teaching students about the influence of social organization on parental care and on its efficiency it is helpful to give information about the organization of social insects, although this is not essential. It may well be enough to compare the organization of a herd and of a family.

Q1 What did you learn through play?

Q2 What did you learn by imitating your parents?

Q3 What did you learn by experience?

Q4 You still have a great deal to learn – what part will you expect parental care to play in this? For how long?

Educative parental care

This and the next section are not suitable for class teaching and are directed rather towards project work and discussion; in fact, this section is probably best suited to small groups, each contributing to a consensus from the whole class on the four questions asked. This could lead to an impressive wall display and even be a possible contribution to closing the 'generation gap'!

Emotional development and parental care in humans

Emotional development depends particularly on

communication skills. Attempts have recently been made in this country to educate parents and potential parents, mainly by directed experience, in the communicative methods and emotional needs of early childhood. Such attempts have included the Denaby Main experiment, where parents were helped to communicate and play with the children in supervised play groups, and the introduction of adolescent girls and boys into play groups where they learn how to give acceptance and security to small children.

Reference material	**Books**

*Reading suitable for students

*Barnett, S. A. (Ed.) (1958) *A century of Darwin.* Heinemann/Mercury Books. (See Chapter 10, Smith, J. M. 'Sexual selection'.)
Campbell, B. G. (1967) *Human evolution.* Heinemann.
(This outlines the scientifically reliable evidence of the biological nature of man's behaviour and its primate origins.)
*Darlington, C. D. (1964) *Genetics and man.* Allen & Unwin.
(Contains stimulating and, at times, controversial discussions of human sexual selection.)
Eibl-Eibesfeldt, I. (1971) *Love and hate.* Methuen.
Gould, J. (Ed.) (1968) *The prevention of damaging stress in children.* (Report of U.K. study group No. 1 to the World Federation for Mental Health.) Churchill.
Green, T. L. (1955) *The teaching and learning of biology.* 2nd edition. Allman. (See Chapter XX, 'Sylviculture as a school biology topic'. This describes methods of artificially propagating trees and shrubs.)
Peterson, R. T. (1968) Life Nature Library *The birds.* Time–Life.
Schools Council Moral Education Project (1972) *Lifeline* series. (A pack.) Longman.
Seglow, J., *et al.* (1972) *Growing up adopted.* National Foundation for Educational Research.
Schools Council Humanities Project (1970) *Relationships between the sexes.* (A pack.) Heinemann.
Stern, C. (1960) *Principles of human genetics.* W. H. Freeman.
(Contains a great deal of information on selection in human populations.)
Tinbergen, N. (1966) Life Nature Library *Animal behaviour.* Time–Life.
World Health Organisation (1962) *Deprivation of maternal care.* World Health Organisation. (A critical appreciation of earlier work. Now out of print but available in large libraries and through inter-library loan.)

Article

*Tinbergen, N. (1952) 'The curious behaviour of the stickleback'. *Scientific American* Offprint No. 414.

Films

'Behaviour of the three-spined stickleback'. Sound, colour, 15 minutes. Oxford Scientific Films, available from G. H. Thompson, Oxford Scientific Films Ltd, Long Hanborough, Oxford OX7 2LD.

Three films in the series, Biology of spiders, each film sound, colour, 10 minutes:
'Comparative courtship' (in 7 unrelated species)
'Comparative courtship in Lycosidae (wolf spiders)'

'Comparative courtship in Salticidae (jumpers)'
Oxford Scientific Films, available from G. H. Thompson, Oxford
Scientific Films Ltd, Long Hanborough, Oxford OX7 2LD.

Two films from the series, The Family of Man:
  'Married life'
  'Teenagers'.
Each film sound, black and white or colour, 50 minutes.
For hire: BBC TV Enterprises Film Library, 25 The Burroughs, Hendon.
For purchase: Non-theatric Film Sales, BBC Enterprises, Villiers
House, Haven Green, London W5 2PA.

Film loops

'The engagement', from the series, Points of Departure. Eothen Films
(International) Ltd, 103–9 Wardour St, London W1V 4PJ.
Nuffield Secondary Science 'Parental care in wild animals' 1 and 2.
0 582 24316 5 and 0 582 24317 3. Longman.

Outline of the work on population control
(Chapter 8)

Introduction

8.1	**The world problem**

Development

8.2	**Ways of limiting population growth**

8.21	Historical

8.22	Limiting population by individual responsibility

8.23	Methods of birth control

Methods concerned with the male reproductive organs :
Castration ; vasectomy ; chemical means ; mechanical means ; withdrawal

Methods concerned with the female reproductive organs :
Ovariectomy ; tubal ligation ; chemical means ; mechanical means ; the 'safe' period ; the I.U.D. ; prostaglandins ; abortion

8.3	**Moral aspects of population control**

Summary
to be made of the work on population control

Background reading	**Genetic counselling**

8
The control of human breeding

This section should encourage students to discuss whether we should be content to allow the old means of population control, war, famine, and disease, to continue or whether we have any more humane alternatives. Whatever new methods of food production are provided, there is obviously a limit to the population the planet can support and in this section the problem is posed.

Introduction
(Section 8.1)

8.1 The world problem

Discussion of the factors affecting world population in the simple scheme provided will need a consideration of what is meant by the word 'moral'. Teachers may prefer to deal with this work in a team teaching session, with a member of staff who is an expert in this field, or to ensure that the class has previously done work on this, possibly by using the material of the Schools Council Moral Education Project, published as the *Lifeline* series of booklets. (See page 139.)

The topic book of the Schools Council Integrated Science Project, *Population patterns* (see page 139), is recommended, possibly for homework once or twice. It would be best if small groups examined discussion questions first and each group then elected a spokesman to report to the rest of the class on selected questions. If a whole class discussion is attempted, the session may be emotional and not logical or critical.

Table 16 in the *Text* shows the birth rate in England and Wales between 1930 and 1970.

Q1 What factors may have influenced the fall in birth rate as indicated by the figures for women in the reproductive ages between 15 and 44?
a In 1935?
b In 1966–70?

a Students may not know that in 1935 in this country, serious unemployment and starvation may have affected the birthrate.
b In 1966–70, a large number of the population discovered the contraceptive pill and began to use it.

Q2 What factors may have influenced the rise in birth rate?
a From 1945–50?
b From 1955–66?

a After the Second World War, factors may have included: the feeling that this was a better time to have children than the previous years; the return of many servicemen to wives they had not seen for two or more years.

b The extra population produced in the post-war 'bulge' was entering the reproductive ages during this period, but since the figures given are per 1000 women, one cannot attribute this rise in *rate* to the 'bulge' children, although this undoubtedly produced an *overall* increase in numbers. The probable reason for the rise in rate was that this was the time known by the slogan 'You've never had it so good'. Birth rates have always risen with affluence, within certain constraints.

Students are asked to discuss questions 3 to 7.

Q3 What factors can be described as 'death control' methods?

Q4 If the result of death control without birth control is a lowering of the quality of life (for example, in Calcutta) can you suggest solutions which are practical both morally and financially?

Q5 What do you mean by 'moral'? Do you, for example, think that behaviour is always right or wrong, or does it come in varying shades of grey? Is abortion always wrong because it involves taking a human life?

Q6 Famine, epidemics, and war have been the major population controls in the past. The world can only supply food for a certain number of people, so when food becomes inadequate because of rising population, these natural controls may take over again. What are the alternatives?

Q7 'The crux of the matter is not only whether the human species will survive, but even more whether it can survive without falling into a state of worthless existence.' (*The Executive Committee of the Club of Rome* in 'The limits to growth'.)

There are no exact answers to these. Students might like to list all the points involved in each question and attempt to produce a class consensus or possibly two, one from the optimists who think that science will provide the answers to food production in time, and one from the pessimists who don't care very much for life as it is today and would certainly not tolerate any worsening of its quality.

8.2 Ways of limiting population growth

1 to allow students to become aware of the powerful mechanisms involved in the perpetuation of the human species

2 to discuss some of the irrational causes of overpopulation and unwanted children

8.21   Historical

1 to show that man has always been aware of population problems, and to show the ways in which he has tried to deal with them

2 to apply and so reinforce the scheme of factors affecting population growth given in 8.1

Q1 Which of the following factors would have had the most influence on an increase in population?
a Jenner's discovery of the smallpox vaccine.
b The organization of public health services; for example, the provision of clean water, which reduced the incidence of cholera.
c Better agricultural methods and improved transport and distribution of food.
d Increased cultivation of the new crops, such as potatoes, from America.

Primitive man was fairly callous about survival and methods of population control used in the eighteenth century seem barbarous today. Teachers may be disappointed that Malthus is not mentioned either in the historical content or in relation to the world problem – he is met with in so many texts it was decided to show that others were also aware of the problem at different times in history.

This sub-section also helps to consolidate the work on public health in *Text* 1, *Introducing living things*, Chapter 8.

This question is for discussion.
The factor not mentioned which contributed to population growth in the eighteenth century was the absence of major wars.

8.22 Limiting population by individual responsibility

Abstinence has always been a method of population control, but not a very realistic one. In the face of all the mechanisms involved in getting male and female together and ensuring that the dangerous process of birth and the hard work of child rearing continue, in a species which is able to think rationally about such things, abstinence is physiologically difficult.

Sometimes conscious choice is made irrationally, due to cultural conditioning; the film recommended, 'My brother's children' (see the Reference material on page 139), shows clearly the problems when a culture changes from being an agricultural one to one which is urban centred. It is a well made film and the drum music of the Yoruba and the colour are well worth hearing and seeing.

Table 17 in the *Text* gives statistics of illegitimacy.

Q2 Are there any rational reasons for this rise?

Q3 Is there any rational reason for a girl under 16 to have a baby?

Q4 What do you think might be the irrational causes? There were 1781 schoolgirl abortions in 1970 and 2646 in 1971.

Again, there are no answers to the problems, though the following facts might be borne in mind:
a that adolescents who experiment with sex at an early age are often those who experiment early with cigarettes, alcohol, and other drugs, often as a compensation for lack of parental care.
b adolescents are unaware that 'adolescent sterility' often saves them from unwanted pregnancy, so giving them a false sense of security, and that it tends to stop with alarming suddenness.

8.23 Methods of birth control

Objectives

1 to give information about the methods of contraception, including the risks involved and the reliability and side effects of each method

2 to enable some evaluation to be made of differing needs for contraception, due to human variation in physical and cultural aspects

Before beginning this work teachers might like to revise the work on the menstrual cycle in *Text* 1, *Introducing living things*. The work of the present chapter assumes that they have done this and takes it further.

Teachers might also introduce the work by mentioning the fact that some humans are infertile, for a variety of reasons, and that some may never become parents in spite of a great desire to have children of their own.

Contraceptive methods are described at some length partly because some pupils may never again have the opportunity to discuss these problems with an informed person. In addition, with influences from the mass media, students cannot fail to hear about contraception and to be anxious about it. A common worry among responsible boys is that there is no Pill for men as yet. Some feel that it is unfair that the female should have all the worry about reliable contraceptives and risk the dangers of the Pill, others worry about tales they hear of girls who deliberately get pregnant to force a man into marriage, by pretending to use contraceptives when in fact they do not. Again this may be an occasion for team teaching.

The ways both the Pill and the I.U.D. work, have, by recent research, been revealed to be much more complex than was at first suspected, so explanations of these devices have been kept very simple indeed; full descriptions of present knowledge are too complex for students at this stage.

The Pill chart in figure 142 in the *Text* shows high oestrogen content to be related to the advance of thrombosis, but there are two anomalous cases.

Table 18 in the *Text* gives statistics of the causes of women's deaths.

Q5 What other valid conclusions can you draw from the figures in *table 18*?

a That the best time to have a baby is before the age of 35.
b That having a baby at any age is a risky business and is not to be undertaken lightly.
c That older women run a greater risk of dying from all causes other than car accidents.

The *Text* discusses the 'safe' period and its disadvantages. These are, chiefly, that both egg and sperm have been shown to live longer than the times given. Also, the egg is not necessarily swept out by the menstrual flow. Taking the temperature to find when ovulation is finished, and abstaining until then, has psychological disadvantages.

Q6 Can you think of other reasons why the 'temperature method' should be unsatisfactory?

a Temperature rise may be due to a mild infection.
b Many people have difficulty in reading a clinical thermometer, and in particular of making readings accurately when dealing with such a small rise in temperature.

Table 19 in the *Text* compares different methods of contraception.

Q7 Which of these methods is the safest?

These questions are for discussion.

Q8 Which is the most reliable?

Q9 Which is the simplest?

Q10 Which method do you think would be preferred by young married people in Britain today?

Q11 Which would be preferred by older married people who already have enough children?

Q12 Which method would be best in an underdeveloped country?

8.3 Moral aspects of population control

Objectives

1 to help students to appreciate the many different valid, if opposing, points of view on the rights and wrongs of abortion and contraception

2 to give students some practice in making judgments when emotions and unalterable previous conditioning may confuse the facts

Discussion on moral issues needs a clear frame of reference for the students on what is meant by 'morals' as opposed to 'moralizing'. Where classes contain students of very different religious beliefs, it would perhaps be best to avoid this section altogether, as logical discussion might well be impossible in such a class situation. Team teaching could be used to advantage.

Summary

At this point students should be asked to make their own summary of the work of this chapter.

Genetic counselling

Background reading

Q1 Colour-blindness is inherited in the same manner as haemophilia but is harmless. The colour picture shown in *plate 10* is sometimes used as one of a series of tests to indicate whether a person is colour-blind.
a What do you see in the picture?

A teapot if you are not colour-blind.
A mug with a handle if you are red–green colour-blind.

b Devise a scheme to show how colour-blindness is inherited.

Students might be expected to produce a scheme similar to figure 147 in the *Text*, but using different symbols for normal colour vision and colour-blind vision. Colour-blindness is recessive and sex-linked but, unlike haemophilia, the double recessive condition (found in ♀ only, of course) is not lethal.

c Can you now explain why girls are rarely colour-blind?

As colour-blindness is sex-linked and recessive, a boy must have either normal vision (N) or be colour-blind (n), depending on which gene is carried on his *single* X chromosome.
Girls have *two* X chromosomes and will therefore only be colour-blind in those rare cases where both X chromosomes carry the n gene. Colour-blind girls must have a colour-blind father and a mother who is either colour-blind or a carrier.

d If half the sons of the daughter of a colour-blind man are colour-blind, explain the inheritance of this condition through the generations.

The simplest explanation of this problem is as shown in *figure 26*, and assumes that the original colour-blind man is the only source of colour-blind (n) genes.

Figure 26

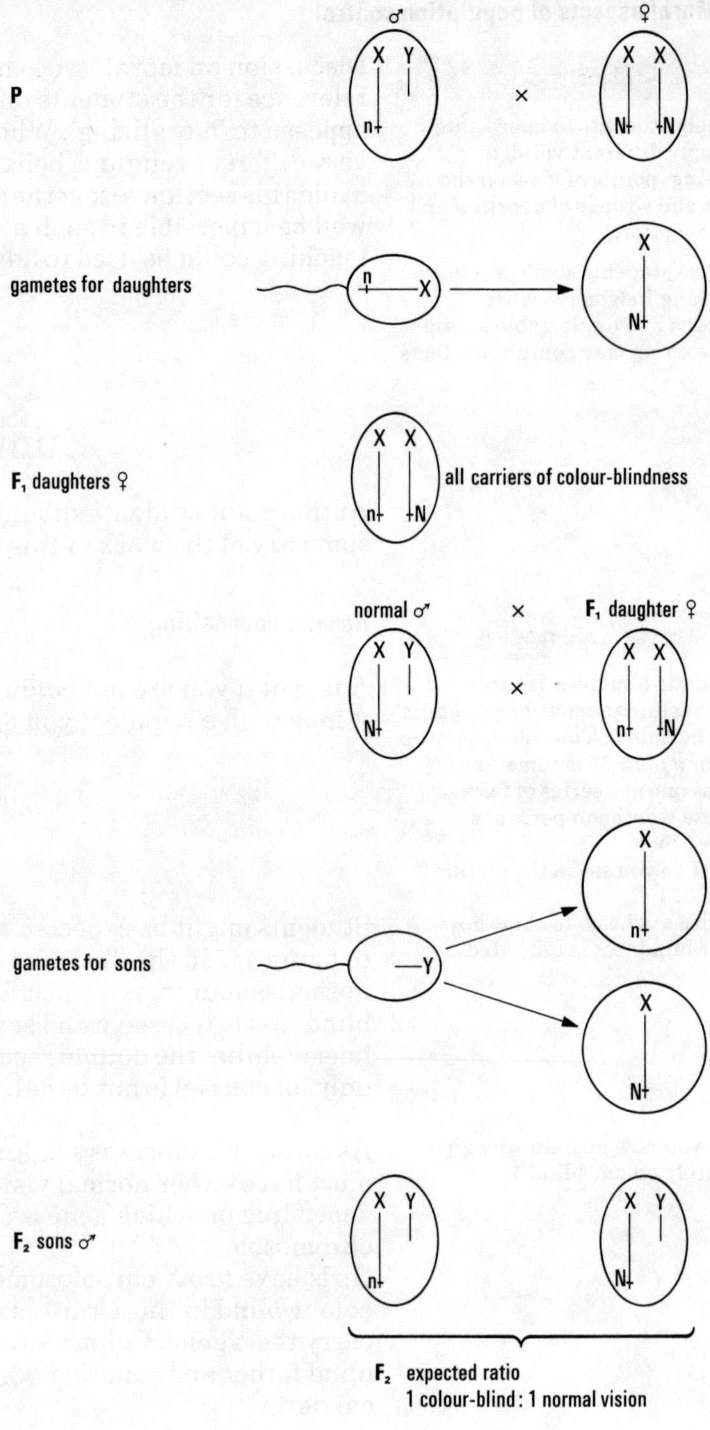

P
♂
♀
X Y
X X
n N N
×
gametes for daughters
n X
X
N
F₁ daughters ♀
X X
n N
all carriers of colour-blindness
normal ♂
F₁ daughter ♀
×
X Y
N
X X
n N
gametes for sons
Y
X
X
N
F₂ sons ♂
X Y
n
X Y
N
F₂ expected ratio
1 colour-blind : 1 normal vision

Q2 Should parents have the right to decide to terminate a pregnancy where it is known that the developing foetus has inherited some 'major handicap'?

For discussion.

Q3 Who is to decide what constitutes a 'major handicap' (thus allowing an abortion)? Does this category include gross mental retardation, slight or moderate mental retardation, blindness, lack of one or more toes or fingers, lack of one or more limbs?

For discussion.

Reference material

The International Planned Parenthood Federation has excellent books, pamphlets, films, etc. at 18 Lower Regent Street, London SW1.

Books

*Reading suitable for students
Burke, S. (Ed.) (1970) *Responsible parenthood and sex education*. International Planned Parenthood Federation.
Club of Rome (1972) *The limits to growth*.
Dallas, D. M. (1972) *Sex education in school and society*. National Foundation for Educational Research.
Dalton, K. (1969) *The menstrual cycle*. Penguin.
Lawton, D. (Ed.) (1971) *Population education and the younger generation*. International Planned Parenthood Federation.
Nuffield Secondary Science (1971) *Theme 3 Biology of man*. Longman.
Peel, J. (Chairman) (1972) *Unplanned pregnancy*. Report of the Working Party of the Royal College of Obstetricians and Gynaecologists. Available from R.C.O.G., 27 Sussex Place, London NW1 4RG.
Peel, J., and Potts, M. (1969) *Textbook of contraceptive practice*. Cambridge University Press.
*Schools Council Integrated Science Project (1973) *Population patterns*. Longman/Penguin.
Schools Council Moral Education Project (1972) *Lifeline* series. Longman.
Wrage, K. H. (1969) *Man and woman: the basis of their relationship*. Collins.

Articles

The following offprints from *Scientific American* are recommended:
Langer, W. L. (1972) 'Checks of population growth 1750–1850'. No. 674.
Pike, J. E. (1971) 'Prostaglandins'. No. 1235.
Tietze, C., and Lewit, S. (1969) 'Abortion'. No. 1129.

Films

'My brother's children'. Sound, colour, 47 minutes. (This illustrates cultural changes among the Yoruba people and the problems of communicating contraceptive knowledge.) Hire from Concord Films.

Film loops

Nuffield Secondary Science 'Some solutions to world food problems'. 0 582 24340 8. Longman.
Nuffield Secondary Science 'World food problems'. 0 582 24339 4. Longman.

Outline of the work on selection
(Chapter 9)

Introduction

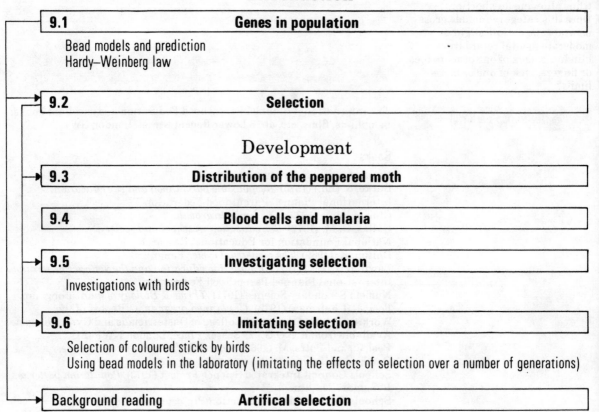

9.1	Genes in population

Bead models and prediction
Hardy–Weinberg law

9.2	Selection

Development

9.3	Distribution of the peppered moth

9.4	Blood cells and malaria

9.5	Investigating selection

Investigations with birds

9.6	Imitating selection

Selection of coloured sticks by birds
Using bead models in the laboratory (imitating the effects of selection over a number of generations)

Background reading	Artifical selection

The selection of a new generation

<table>
<tr><td>

Objectives of this chapter

1 to introduce the idea that a study of genes in populations is important

2 to develop the concepts of genotype and phenotype

3 to provide an opportunity to follow up population genetics with bead models and the Hardy–Weinberg equation

4 to present the idea of selection in a population and provide evidence of selection

5 to encourage students to design and carry out experiments to investigate selection

6 to use models to imitate selection and demonstrate *a* that camouflage has survival value, *b* that selection over a number of generations will have profound effects on the proportion of genotypes in the population

7 to develop an understanding of the idea of a 'gene pool' and of the effects of selection on this pool

</td><td>

Using the term 'evolution'

Up to this point in the course the term 'evolution' has been carefully avoided. It appears that many students have preconceptions, often strongly held, about evolution and tend to consider it as something to do exclusively with the past. To them, evolution is just about fossils, the origin of life, Genesis, and Charles Darwin!

One fundamental aim of this course is to outline the *mechanism* of evolution as it is occurring *at the moment*. This allows students to deal with more tangible material and provides them with a sound base from which they can extend their studies into the problems of past and future evolution. Such topics are considered in Chapter 10.

It is possible to deal with the topics of Chapter 9 without deliberately using the term evolution and teachers are strongly recommended to do this. In this way the adverse effects of preconceptions can be avoided – to some extent, at least.

</td></tr>
</table>

Introduction
(Sections 9.1 and 9.2)

9.1 Genes in populations

To introduce the idea that a study of genes in populations is important

Objective

Q1 Can you roll your tongue (see *figure 149*)?

Q2 Can other members of your family roll their tongues? Can everybody in your class?

This section has been kept brief and simple in the *Text*. Teachers may choose to develop the theme with the 'Additional work on genes in populations' (page 142), or move on to the work on selection. This decision will be governed by the interest of students and their ability to deal with mathematical concepts. If teachers prefer, they could postpone this additional work and do it in conjunction with section 9.6 'Imitating selection'.

Yes or no!

This dual question is to encourage students
a to investigate the principle governing the inheritance of tongue rolling
b to collect data which will develop the idea of a 'gene pool' in a population.
If any problems arise in explaining inheritance in a family,

it should be pointed out that some people have developed the ability (consciously or unconsciously?) to roll their tongues.

Q3 Write down the letters which represent the genotypes (genes carried) of rollers and non-rollers.

Rollers: RR (homozygous), Rr (heterozygous) non-rollers: rr

Q4 Which phenotype (roller or non-roller) has *two* genotypes?

Roller.

Now follows a problem on the inheritance of a hypothetical recessive human disease.

Q5 What is the probability that you, a fit person, are a 'carrier' (that is, Dd)?

No one is expected to *know* the answer to this question so students are told that it is about 1 in 16. It is, of course, calculated by using the Hardy–Weinberg equation (see page 149).

Q6 What is the probability that you (Dd) will marry somebody who is also a carrier (Dd)?

About 1 in 16.

a 1 in 4; *b* 1 in 2; *c* 1 in 4.

Q7 If you (Dd) do marry somebody who is a carrier, what is the chance that your first child will
a suffer from the harmful disease?
b be a carrier of the disease?
c be perfectly normal?

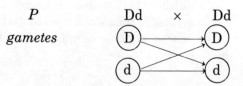

P   Dd   ×   Dd
gametes

progeny ratio: DD: Dd Dd: dd

Q8 What is the chance that your second child will suffer from the disease?

Still $\frac{1}{4}$! It does *not* matter what the first child was: normal or sufferer.

Q9 What is the chance that a normal married couple will produce a child who has the disease?

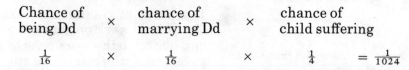

Chance of being Dd	×	chance of marrying Dd	×	chance of child suffering	
$\frac{1}{16}$	×	$\frac{1}{16}$	×	$\frac{1}{4}$	$= \frac{1}{1024}$

This last question gives teachers the opportunity to point out that it is mainly normal people (carriers) who produce the 1 in 1000 child who suffers from the inherited disease.

Additional work on genes in populations

One or both of the sections below can now be introduced to develop the theme.

The perpetuation of life

Apparatus and materials
Per group:
200 beads of one colour (e.g. *red*)
200 beads of another colour (e.g. yellow)
5 containers for beads (beakers, small cardboard boxes, etc.)

Six ideas need to be clear in the student's mind in order to comprehend the work of this section:

1   That we are only concerned with the inheritance of pairs of characteristics of which an organism may have one or the other, but not both. These are termed alleles, and we can also call the gene responsible for such characteristics an allele. The term allele as used here refers to a mutually exclusive relationship between two or more genes and between the characteristics for which they are responsible. Originally this term referred to a characteristic rather than to the gene responsible for it. Usage has gradually changed and nowadays it is frequently intended to refer to alternate genes found at a particular locus in a chromosome. In fact, allele can be used in either sense without confusion.

2   That we are only concerned in this work with a gene pool containing two alleles.

3   That we are concerned with the genes of the zygote or a single cell of an organism.
    The cells of a multi-cellular organism will contain, altogether, an enormous number of chromosomes and, hence, genes. However, assuming development occurs through mitosis, each cell will have the same number of the same type of genes. Thus, the organism can be *represented* by the genes of one cell which will be the same as the original fertilized egg or zygote. The genes in the zygote make up the genotype of the organism.

4   That the gene pool of a population contains those genes that can contribute to the genotypes of future generations. They are the genes that, through segregation, are present in the gametes.

5   That in this work we are dealing with a population breeding at random. In Chapter 2, when beads were used to demonstrate the inheritance of characteristics, only one cross was possible in each case studied. In other words, these were examples of populations not breeding at random. It is worth pointing this distinction out to the students.

6   That fertilization is random irrespective of whether breeding, that is, mating, is random or non-random.

The following work schedule will need to be provided for the students.

When two organisms are mated in the laboratory it is easy to predict what their offspring will be like if we know the genotypes of the parents. If, for example, one is a pure-bred, ebony-bodied *Drosophila* the other a pure-bred *Drosophila* of normal body colour we know the offspring will all have normal bodies.

However, suppose you have not one pair of organisms but several, forming a population and inter-breeding at random. This is the usual situation under natural conditions and its analysis is a little more complex. In a population of, say, 100 pure-bred *Drosophila* of normal body and 100 pure-bred ebony-bodied *Drosophila* with the sexes distributed equally, three types of matings are possible.

1 Ebony-bodied flies can mate with each other.
2 Flies of normal body can mate with each other.
3 Ebony-bodied flies can mate with flies of normal body.

We are assuming that mutation does not occur and that mating between the flies is random.

How can we predict the characteristics of the offspring in such a mixed population?

When mating takes place the act of fertilization brings genes from each of the parents together. These genes determine the characteristics of the offspring. When an ebony-bodied fly mates with a fly of normal body it is the coming together of the gene for ebony (e) and that for normal body (+) which causes the offspring to be hybrids and to have normal body colour.

When studying problems of mating in populations it is important to remember that we are concerned with *genes not whole organisms*. Provided we remember this, analysis is really quite easy.

In the mixed population of *Drosophila* mentioned above we had 100 pure-bred flies of normal body colour.

We know that two genes (alleles) responsible for the characteristics of normal body were present in the fertilized egg (zygote) that gave rise to each fly.

When such a fly forms gametes, each contains one member of each pair of genes. Thus, although many gametes are produced, they are of two kinds only. One kind will carry one member of a pair of genes; the other will carry the second gene of the pair.

Thus, for a particular characteristic, we can represent the genotype of the organism by a pair of genes and each of its gametes by one member of that pair.

In our population there are 100 pure-bred flies of normal body represented by $100 \times 2 = 200$ genes responsible for the characteristic of normal body. This will, in fact, be the number in the zygotes from which the flies developed.

Likewise, the 100 pure-bred ebony-bodied flies can be represented by $100 \times 2 = 200$ genes responsible for the development of ebony body.

If we denote the gene for normal body as + and that for ebony body as e then the total number of genes that contribute to the formation of either normal body or ebony body in this population can be represented as 200e and 200 +.

It is the total store of genes in a population (the gene pool) that will give rise to the characteristics of the offspring. This is what we must study if we are to predict the composition of a future population (*figure 27*).

We can construct a simple model of a gene pool by representing different types of genes with beads of different colours. In our fly population we can represent a + gene by a red bead and an e gene by a yellow bead. 200 red beads and 200 yellow beads will together provide a model of the gene pool.

We can find what happens at fertilization in this way. Assuming there are equal numbers of males and females, there will be 100 e genes in eggs and 100 e genes in sperms. Similarly 100 + genes will be in eggs and the same number of + genes in sperms.

1   Place 100 red beads and 100 yellow beads in one container and the same numbers in another. The contents of the first container can represent the males and the beads the genes in the sperms. In the second container are the females, the beads representing the genes in the eggs.

2   Shake the beads in each container so that they are thoroughly mixed. Then, without looking, pick one bead out of each container. In doing this you will have imitated the random coming together at fertilization of a gene in a sperm and a gene in an egg.

3   Each pair of beads represents the genotype of one of the offspring. Place the pairs of beads you draw out together. When you have used up all the beads (that is, finished all the fertilizing) add up the number of each of the three genotypes.

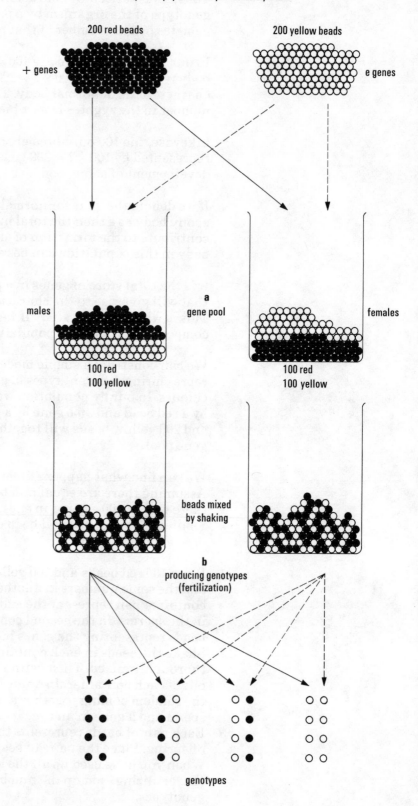

A parent population represented by beads

200 red beads   200 yellow beads

+ genes   e genes

males   **a**   females
               gene pool

100 red   100 red
100 yellow   100 yellow

beads mixed
by shaking

b
producing genotypes
(fertilization)

genotypes

Figure 27
Using beads to represent **a** a gene
pool and **b** the production of
genotypes in a population. The
arrows represent the ways in
which the beads are transferred
into and out of the containers.

Here are two hypotheses that might account for the results you obtain:

First hypothesis
There are three possible matings:
1 ebony body × ebony body
2 normal body × normal body
3 normal body × ebony body

We can represent these matings as follows:
1 e × e = ee
2 + × + = + +
3 + × e = +e

Thus, it might be expected that the proportions of the genotypes in the offspring would be: 1 ee: 1+ +: 1+e

Second hypothesis
Each e gene in the population can either be associated with another e gene or a + gene at fertilization. Similarly each + gene in the population can either be associated with another + gene or an e gene at fertilization. These possibilities can be expressed as follows:

Male gametes

		e	+
Female gametes	e	ee	+e
	+	+e	+ +

If we add up these possible gene combinations you will see we can expect 1 ee : 2 + e : 1 + +.

Students can now consider the following:
a Which hypothesis fits the results obtained from your bead model?
b Does the bead model help you to understand why the particular hypothesis is correct?

Let us now continue our breeding experiment. What would be the make-up of the genotypes in the next generation? Try to find an answer, using the beads in the following way.

We have assumed in our population that the numbers of the two sexes are equal. Thus, we can assume half the numbers of the organisms of each genotype to be males and the other half to be females.

Place half the number of beads representing each genotype in the container for males and half in the one for females.

Shake each container thoroughly and then take out a pair of beads at random, one at a time, from each container. Score the genotypes just as you did previously.

Repeat this process several times. On each occasion you will be producing the equivalent of a new generation.

Questions to consider are:

c Do the proportions, or *frequency*, of each genotype vary from one generation to the next or remain constant?

d When you have completed your breeding with the bead model try to construct a hypothesis to account for what happens over several generations in a gene pool such as you have imitated.

Analysis of results
In answer to question *a* above, the genotypes obtained will be in the proportions 1 ee: 2 +e: 1 + +
or, as beads,
1 yellow yellow : 2 red yellow : 1 red red
This agrees with the second hypothesis.

In answer to question *b* above, the first suggestion is obtained by considering the organism, not the gametes. The second suggestion is worked out by considering the gametes.

It is the gametes produced by the organism, not the organisms themselves, which are directly responsible for producing the genotypes of a new generation.

As gametes are produced in pairs, each containing one of each pair of alleles, a member of a pair has the possibility of being associated with *either* of the members of the other pair in fertilization. This point is taken into account in the second suggestion but not the first.

The beads are used to represent genes carried by the gametes and thus represent the process accurately. The results obtained by the bead model confirm the correctness of the second hypothesis.

In answer to question *c* above, the proportions, or frequency, of each genotype remain constant.

In answer to question *d*, gametes of each generation will contain the alleles in the proportions in which they are found in the genotypes. Thus, in the formation of each generation the same proportions of alleles are to be found in the gene pool. As the gametes fertilize at random the same combinations are possible in the production of each generation. It follows that the proportions of the genotypes in each successive generation must remain constant.

Using the bead model in association with work on selection
The results obtained with the bead model are those you would expect from a population which is free from the influence of selection. Therefore they can be used as a basis for comparison in order to measure the amount of selection acting on a similar population. This is considered further in 9.6.

The Hardy–Weinberg law

According to the Hardy–Weinberg law, the genetic composition of successive populations cannot change, and the frequencies with which the three possible genotypes of a single gene occur in each population remain constant, so that these frequencies are:

frequency of homozygous dominant genotype $= p^2$
frequency of heterozygous genotype $\quad = 2\,pq$
frequency of homozygous recessive genotype $= q^2$

where p is the frequency of the dominant allele A, and q is the frequency of the recessive allele a and where $p+q=1$, when $p=1-q$.

That this must be the case can be presented in a single figure, in which the types of gametes in a population and their frequencies are represented by the axes of a lattice (see *figure 28*) and the possible combinations amongst the offspring, together with their frequencies, by the co-ordinates. This figure is built up in three stages.

1 Identify the range of possible genotypes amongst the offspring when, in the parental generation, gametes carrying the dominant and the recessive alleles of an arbitrary gene A are distributed amongst both sexes:

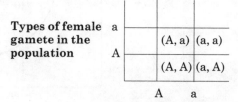

Types of female gamete in the population

<table>
<tr><td></td><td>a</td><td>(A, a)</td><td>(a, a)</td></tr>
<tr><td></td><td>A</td><td>(A, A)</td><td>(a, A)</td></tr>
<tr><td></td><td></td><td>A</td><td>a</td></tr>
</table>

Types of male gamete in the population

Genotypes amongst the offspring: AA, Aa, aa.

2 Introduce the frequencies with which the types of gametes occur.

Suppose alleles A and a are equally distributed amongst the gametes of both sexes and their frequency is unknown. Let the frequency of allele A be p and the frequency of allele a be q. By introducing these frequencies into the lattice, the frequencies of the genotypes in the next generation are immediately apparent.

		A (p)	a (q)
Types and frequency of female gametes in the population	a (q)	(A, a) pq	(a, a) q^2
	A (p)	(A, A) p^2	(a, A) qp

A (p)   a (q)

Types and frequency of male gametes in the population

Frequencies of genotypes amongst the offspring are:
$$AA = p^2$$
$$Aa = 2\,pq$$
$$aa = q^2$$

3 Introduce the idea of successive generations.
The offspring themselves eventually reach maturity when, according to the Hardy–Weinberg model, they will form gametes in the manner shown in *figure 28*, demonstrating that, within the parameters of the Hardy–Weinberg law,

Figure 28
A diagram summarizing the Hardy–Weinberg law.
From Dudley, B. A. C. (1973) 'The mathematical basis of population genetics', Journal of biological education, *7*, 41–3.

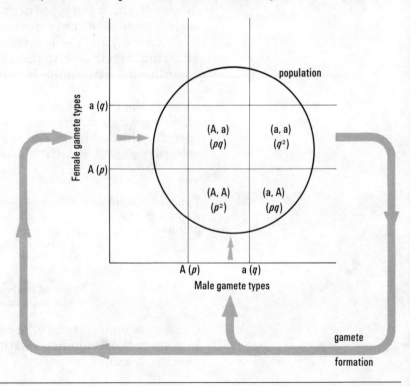

The perpetuation of life

there can be no further change in the genetic composition of the populations of successive generations. Since the frequency, in a population of the genotype aa, can be established by counting the number of individuals showing the recessive phenotype in a population of known size, then q^2 is known and q can be calculated. By means of the model both p and the frequencies of the remaining genotypes of the gene concerned also can be established, giving the genetic composition of the population with respect to that gene.

According to Dudley (1972), if the incidence of the homozygous recessive condition is b, a decimal fraction, then the incidence of the homozygous dominant condition is $1 - 2\sqrt{b} + b$ and the incidence of the heterozygous condition is $2(\sqrt{b} - b)$. This simplifies the calculation of frequencies of, for instance, the carriers of a number of different congenital conditions in humans and hence the likely number of carriers in a population of known size. A number of worked examples and further details for practical investigations of the Hardy–Weinberg law are contained in the same reference.

Suggested exercises

Tongue rolling   Collect data of rollers and non-rollers for a large number of people (for example, the whole school) and then use the Hardy–Weinberg law to predict the proportions of the genotypes.

Tasters and non-tasters   The ability to taste phenylthio-carbamide (PTC) is controlled by a single dominant gene (T). An investigation amongst school and families could identify the tasters (TT, and Tt) and non-tasters (tt).

Phenylthiocarbamide (PTC) is a dangerous chemical. Students should not be allowed to handle the solid or dilute solution. They should only be provided with strips of filter paper impregnated with 0.13 per cent PTC as indicated in Appendix 1. Plain strips of filter paper can be used as a control in this experiment.

Q5 What is the probability that you, a fit person, are a 'carrier' (that is, Dd)?

The problem set in 9.1, question 5   This assumes that there is a recessive gene (d) responsible for a fairly serious disease in humans. 1 in 1000 of the population suffer from the disease.

From Hardy–Weinberg:

$$\underbrace{p^2}_{\text{DD}} + \underbrace{2pq}_{\text{Dd}} + \underbrace{q^2}_{\text{dd}} = 1$$

$$0.999 \quad + \quad 0.001 = 1$$

also $p + q = 1$ where p = frequency of gene D
q = frequency of gene d

$$q^2 = 0.001$$
$$q = \sqrt{0.001} = 0.032$$
$$p = 1 - q$$
$$p = 0.968$$

$\therefore$ *either*, $2\,pq = 2 \times 0.968 \times 0.032 = 0.062$

or, $p^2 = 0.968^2 = 0.937$

$$q^2 = 0.001$$

$\therefore 2\,pq = 1.000 - 0.938 = 0.062$

$\therefore 0.062$ or 6.2% is Dd

6.2% is equivalent to 1 in 16.1 people.

Approximately 1 in 16 people are carriers.

9.2 Selection

Objective

To present the idea of selection in a population

The concept of selection is based on the influence of environmental factors in favouring the reproduction of certain alleles or combinations of genes over others present in a population.

It is important to keep before the minds of the students the idea that although selection frequently influences the *phenotype*, it is the *genotype* that determines the composition of a future generation. Thus, for example, heterozygotes selected through a dominant characteristic will pass on a recessive gene that can contribute to a future phenotype previously eliminated by selection. A clear understanding of the roles of phenotype and genotype is fundamental to the work of this chapter.

The work of Chapter 7 leads to the idea of sexual selection, which is then brought out in this section. Sexual selection has been deliberately mentioned because it provides an example of selection that does not usually involve death. It is important to produce a balanced picture of selection from the beginning. The idea of selective agents always being lethal is not only a half-truth but can lead to misconceptions, particularly when artificial selection and selection in human populations are considered.

Sexual selection is used in this section in the sense of the selection involved when an animal chooses a mate. It is one form of selection that influences the union of two particular gametes in an act of fertilization. Akin to, but of course, not strictly comparable with, sexual selection in plants and also in those animals that disperse their gametes freely, are selective agents affecting the transport and survival of gametes, e.g. wind, water currents, and insects. See the Nuffield O-level film loops, 'Pollination by wind' and 'Pollination by insects', which illustrate this point. (Details are on page 168.)

Development
(Section 9.3 to end of chapter)

9.3 Distribution of the peppered moth

The account in this section is based on the work of Dr H. B. D. Kettlewell. It provides an example of the way the influence of selection can be affected by the relationship between the *phenotype* and the environment.

The light form of *Biston betularia* referred to is *typical*. The term dark form encompasses both *carbonaria* and *insularia*. Both these are melanic forms, *carbonaria* being uniformly dark while *insularia* has a uniformly dark body but its black wings are slightly speckled with white, especially the hind pair. The account, like that of section 9.4, is arranged in stages, each stage ending in a problem. It could be used as the basis for discussion work in class or tackled by the students in homework or preparation. If the latter course is adopted it is suggested that some introductory discussion be held in class in order to avoid unnecessary confusion in the students' minds. The answers to the questions worked out by students might also be discussed profitably in class.

For discussions it is probably best for the illustrations in the *Text* to be displayed to the class as a whole. The pack of diagrams for making overhead transparencies contains copies of the principal illustrations.

Q1 Does a comparison of the two maps suggest to you a hypothesis that might explain the distribution of the differently coloured moths?

The dark form is found mainly in industrial areas. Where it is found outside the cities it is to the east. Thus, as the prevailing wind blows from the south-west, it would appear that the distribution of the dark form is affected in some way by industrial conditions and particularly those features that are carried by the wind. The polluted atmosphere containing soot and other materials is the obvious possibility.
The light form is found in non-industrial conditions and to the west of the industrial towns. It survives where air pollution is absent.
Several hypotheses could be put forward to explain this correlation between pollution and the distribution of the two forms of the speckled moth.

1 Air pollution affects the bodies of the moths and turns them black when they enter industrial conditions. This hypothesis could be tested by breeding light form moths in simulated industrial conditions in order to see if they, or their offsprings, turned dark. The evidence available from investigations of this type indicates that

---

this is not the cause of melanism.

2   Air pollution causes an enormously rapid mutation rate so that new black forms are being continually produced. Table 20 in the *Text* indicates that in some areas the proportion of dark forms is extremely great. This suggests a mutation rate much above any normally encountered (see Chapter 4). Indeed we should have to assume that the mutagenic effect of the materials in the polluted atmosphere is many thousands of times greater than that of the most powerful doses of penetrating radiation which can be used without producing sterility.

3   The bodies of the dark forms are more able to withstand polluted conditions than those of the light forms. This hypothesis could be tested by breeding the two different forms in polluted conditions and measuring their survival. In fact such investigations by Kettlewell have shown that the dark form does display an advantage in this respect. However, there still remains the problem of the relative absence of dark forms in non-industrial conditions. If it is so hardy why does it not survive there?

4   In light coloured surroundings the dark moth would appear conspicuous, while in dark coloured surroundings the light moth would be more easily observed. Thus camouflage might influence their distribution. The more conspicuous moths would be more liable to predation.

In 3 and 4 the distribution of the two forms of moth is considered to be determined by *selection*. In 3 the polluted conditions act as the selective agent, while in 4 selection is a result of the interaction of the polluted conditions and the predators. Thus, although we may consider the predators to be the selective agent, their influence is apparent only because of the influence of the polluted conditions in darkening the surroundings.

Hypothesis 4 is used in the next stage of the account. The experiment described tests this hypothesis. (A test of significance should be applied to the results shown in table 20.)

Q2 In an area, assume that 40 per cent dark and 60 per cent light moths are present. What interpretations of their distribution are possible?

It could be that these were the proportions in which the moths bred. It could also be that they first appeared in different proportions but that one form survived better than the other. For instance, one might be more liable to be eaten by birds.

Q3 How do the proportions in which the moths were released compare with those in which they were recovered?

Approximately twice as many of the light forms (compared with the dark forms) were recaptured in Dorset. The reverse was the case near Birmingham. The differences are significant.

Q4 Do these observations and the results shown in *table 20* bear out your hypothesis explaining the distribution of the differently coloured moths?

In both habitats a significantly greater number of the more camouflaged forms were recaptured compared with the more conspicuous forms. This supports the hypothesis. During discussion it is well to remember that the results in either locality considered alone might be interpreted in four ways.

1 Dark forms might be more attracted to the light of mercury vapour lamps and to virgin females than are the light forms.
2 The life-span of the two forms might be different.
3 The moths might disperse in and around the habitat to different extents.
4 The two forms might be subjected to predators to different extents. However, since the frequency of recovery is almost exactly reversed in the two habitats, *1*, *2*, and *3* would not appear to be tenable.

Q5 Have you now got all the evidence you feel you need to confirm your hypothesis? If not, what investigation would you undertake to complete it?

No. Direct evidence showing that predation actually occurs has not been presented. Observations showing that predators do take the moths in different proportions according to how much they are camouflaged are necessary. The Nuffield O-level film loop 'Selection by predation' showing birds feeding on moths would provide this evidence. (See the Reference material, page 168.) Alternatively, photographs or slides could be used.

Q6 Can you explain the enormous change in the distribution of the light and dark types of *Biston betularia* in Manchester between 1848 and 1895? It may be helpful to refer to a history book.

The growth of industry during the last half of the nineteenth century led to the deposition of soot and other materials on trees and walls in the Manchester area. Previously the light form of *Biston betularia* had a selective advantage and was more numerous. After industry became established the dark forms were camouflaged and the light forms were not. Thus the latter were more frequently eaten by birds and the dark forms survived to produce predominantly black offspring and became more plentiful.

Pupils are told that the proportions of dark moths found in the Manchester area is now significantly less than it was about 20 years ago.

Q7 How do you account for this change?

Industry has modernized its methods and the level of air pollution created by sooty material has reduced considerably during the last 25 years. Consequently the selective disadvantage for lighter forms has fallen. Light forms are surviving to breed in greater quantities.

Practical work

If it is convenient, the investigation 'Selection by birds', section 9.5, can be started here. It could provide evidence of differential predation of *Biston betularia* by birds and could lead to a more detailed study incorporating the investigation in which models are used.

9.4 Blood cells and malaria

This account is based on the work of A. C. Allison. It provides an example that shows the way the influence of selection can be affected by the relationship between the genotype and the environment.

Figure 154 in the *Text* poses the question whether the family tree it shows indicates that a single pair of genes is responsible for the inheritance of the sickle-cell condition.

There are three phenotypes in the family tree, which is a number you could expect if a single pair of genes (alleles) were involved. The offspring of each generation consist of the phenotypes that it is possible for the parents to produce, assuming a pair of genes is involved.

As only a small number of individuals are involved in the family tree, it is important to make the reservation that it may not show all the possible phenotypes that could be produced.

Students are now told that the blood of a group of children was tested to see if there was any relation between the presence of sickle cells and parasites.

Q1 *Table 21* shows the results of this work. What do they tell you?

The results show that there is an inverse relationship between the presence of sickle cells and malaria parasites in the blood of the children tested.
(A test of significance should be applied to the figures.)

Students are told of a further experiment when people with and people without sickle cells were infected with malaria parasites.

Q2 In *table 22* you can see the results of this experiment. Do they indicate that the presence of sickle cells prevents a person from being affected by malaria?

Yes. The results are very conclusive within the sample. However, one would have reservations because of the smallness of the sample.
It is worth while pointing out to the students that the volunteers were all of approximately the same age and physical condition and that those who developed malaria were cured of the disease.

Q3 Look again at the genotypes of people with the sickle-cell condition. Can you work out more precisely what the selective agents influencing the survival of these people are?

The investigations show that, under certain conditions, a heterozygote can have a selective advantage over both homozygotes. People who are homozygous for normal red cells ($Hb^A Hb^A$) are susceptible to malaria, those who are homozygous for sickle cells ($Hb^S Hb^S$) suffer from anaemia.

The heterozygote (Hb^{A}Hb^S) is apparently immune to malaria yet does not suffer from anaemia.

And from this, can you work out the likely proportions of people with the different types of red blood cells in future generations?

In areas where malaria is prevalent, heterozygotes will tend to increase in proportion to either of the homozygotes. Their blood contains the sickle-cell chemical, but not in sufficient quantities to cause anaemia. In areas where malaria is not present the homozygote containing sickle cells will tend to be eliminated. The other homozygote and the heterozygote will tend to increase in proportion. The presence of heterozygotes always makes it possible, assuming random mating, that some sickle-cell homozygotes will be present in a population.

In the investigation 'Selection and sickle-cell anaemia' of section 9.6 (page 165), bead models of selection acting on populations containing the sickle-cell gene are produced. This could be introduced as an extension to this last answer in section 9.4.

9.5 Investigating selection

Objective

To encourage students to design and carry out experiments to investigate selection

It is necessary for the teacher to supervise students to make sure they do not get discouraged, but it is suggested that supervision might take the form of discussion with individual groups about the ideas and techniques they are employing. The students should be encouraged to refer back to the different aspects of investigation and other relevant topics that they have dealt with in earlier parts of the course.

A class practical at this time in the course will consist of a number of small groups of students each concerned with a different investigation. Periodically they will report their findings to the rest of the class for discussion. Thus, everyone is kept informed of the work in progress and can contribute, indirectly at least, to each investigation.

The investigations of this section could be carried out concurrently by different groups in a class and could last as long as facilities allow. Additional investigations of a similar nature could also be performed so as to provide members of a class with a greater choice.

Selection by birds

This investigation can be used with those that follow in section 9.6, 'Imitating selection'.

Seeds
Mixtures of seeds such as those commonly sold for feeding to canaries or pigeons can be used, or a collection of seeds

from wild plants could be obtained. Equal numbers of each type of seed should be used and spread as randomly as possible throughout the habitat. Care must be taken to ensure that the seeds can be easily collected. Thus, the area in which they are distributed should not be too large.

As a check of the observers' ability to collect the seeds after spreading, disperse a known number of seeds in an area. Cover the area so as to prevent predation and after a period of time – say a day – collect the seeds. If this is repeated by all the students in a group, the mean percentage of each type of seed collected can be used as a base from which to measure selection. Thus as a unit of selection of a type of seed, we can use:

Percentage collected of type of seed which had not been exposed to selection minus percentage collected of type of seed which had been exposed to selection.

The time interval between distributing the seeds and collecting them must be kept constant. This will depend on the timetable and facilities available.

The results of preliminary work using a large variety of potential food organisms will allow hypotheses to be made about the characteristics of the organisms that enable the birds to select them. Observations of the birds taking the organisms will provide additional evidence.

Thus it is likely that the nature of the extreme types, that is, those most selected and those most rejected, can be observed. This can lead to experiments in which the birds are presented with a choice of only the two extreme types, and situations in which there is no choice, that is, when just one type of organism is presented at a time. Other relevant information could be obtained by building models such as those suggested in the investigation 'Using models to study selection', portraying the major characteristics of the selected and rejected types. In this way the hypotheses can be confirmed, rejected, or modified.

With interested students these investigations can lead to more sophisticated work. They provide a good start for projects undertaken at sixth-form level. This work can also lead to studies that show how different aspects of biology are inter-related. Thus, observation of the birds may lead to consideration of bird behaviour and psychology. What causes a bird to prefer one thing to another? How much does motive, such as hunger, affect what a bird prefers? Are a bird's preferences learned or innate? Such questions may lead students to a broader consideration of coloration in animals and plants. What are the different methods of camouflage employed by animals? What is the relative effectiveness of cryptic and warning coloration?

Using models to study selection

Models of moths, seeds, and other organisms can be made from raw pastry consisting of a mixture of three and a half parts of flour to one and a half parts of lard. This mixture is of about the same consistency as Plasticine and can be moulded into almost any shape. It is a good food, containing carbohydrate, fat, and vitamin complex, and is very attractive to birds. It is very stable and can be easily coloured with culinary dyes by means of a paint brush. The dyes should be used at concentrations of one in several thousands and only applied superficially to the models. The shape, size, and colour of the models will depend on the organisms they are intended to imitate.

The models can be left in a habitat for a convenient period of time, although it is wise to find out first what species of birds feed on them. Some, for instance, crows and starlings, are more voracious than others. The models can be used to study the effect of both cryptic and warning coloration and shapes. The relative influences of different methods of camouflage could also be investigated if this is relevant.

Five methods of camouflage are considered to be used by animals:

1 Coloration resembling the animal's background.
2 Counter-shading in which the dorsal surface is darker than the ventral surface. (Uniformly coloured objects subjected to lighting from above are framed conspicuously by their dorsal surface appearing lighter than the bottom.)
3 Disruptive coloration in which the contrasting colours of an organism's body apparently break up its outline over a background of similar colour patterns.
4 Shadow elimination.
5 Mimicry in which the animal resembles some other object unattractive to its predators.

In all the investigations of section 9.5 we are predominantly concerned with characteristics of prey which are perceived visually by predators. Other senses can be used for detection of prey. Again, this is a topic that might be followed up by an interested student.

Q1 Which types of organisms are selected by the birds?

Q2 Can you work out a hypothesis to suggest what characteristics of the organisms encourage the birds to select them for food and are therefore a disadvantage for survival?

Q3 Is it shape, size, colour, or some combination of these characteristics that is selected by the birds?

Q4 To what extent does camouflage appear to play a part in selection? Is it always the apparently disguised forms that survive?

Q5 Some organisms appear not to be attacked because their colour acts like a warning signal to would-be predators. The brightly marked red and black scarlet tiger moths (see colour *plate 11*) are good examples. They are far from being camouflaged, yet they survive well. Do any characteristics of your models appear to act as warning signals to birds?

The answers to questions 1 to 5 will vary according to the organisms studied and the depth of treatment.

9.6 Imitating selection

Objectives

To use models to imitate selection and demonstrate *a* that camouflage has survival value *b* that selection over a number of generations will have profound effects on the proportion of genotypes in the population

The two investigations here might be studied concurrently by the class. Two groups of eight could carry out the camouflage experiment while the remainder (in groups of about 3) use the bead model.

Camouflage

This investigation is based on a report by Patterson *et al.*, 1972. (See Reference material, page 167.)

Apparatus and materials
Per group of eight:
100 red toothpicks
100 green toothpicks
An area of grass about 15 m × 15 m
An area of 'brown dirt' about 15 m × 15 m

Toothpicks can be coloured with vegetable dyes or suitable laboratory stains.

Alternative materials
3 cm lengths of green and red plastic milk straws
green-coloured, hard tennis court
playground

The area of grass required depends on the length of the grass and number of 'birds' to be used.

Cocktail sticks are rather hard and sharp if accidentally left on the playing field.

With careful advance trials these experiments will be a success.

The work could be extended to simulate the effectiveness of long beaks (tongs) and short beaks (hand) or feeding in flocks compared with individuals feeding separately.

Results
Table 5 shows a typical set of results from a group.

Habitat: grass	Trials and prey type												
	1		2		3		4		5		6		7	
Birds	R	G	R	G	R	G	R	G	R	G	R	G	R	G
1	12	13	7	8	7	3	1	1	0	2	0	0	0	1
2	9	4	8	5	5	4	0	2	0	1	0	1	0	0
3	5	2	6	3	3	1	3	3	0	1	0	1	0	0
4	4	3	3	4	2	3	0	2	0	0	0	0	0	1
5	4	4	5	3	1	2	0	2	0	1	0	1	0	0
6	7	4	3	5	1	3	4	4	0	1	0	0	0	1
Total	41	30	32	28	19	16	8	14	0	6	0	3	0	3
Cumulative frequency	41	30	73	58	92	74	100	88	–	94	–	97	–	100
Survivors	59	70	27	42	8	26	0	12	0	6	0	3	0	0
Percentage of survivors that are green	–	54	–	61	–	76	–	100	–	100	–	100	–	–

Table 5
Results of a camouflage experiment with red and green toothpicks. *Data modified from Patterson, R., Custer, T., and Brattstrom, B. H. (1972) 'Simulations of natural selection',* American biology teacher.

Q1 Why were records made of *individual* 'birds'? Why not simply add the *total* for the six 'birds' at each trial?

One or more members of the class may be red–green colour-blind (this characteristic is sex-linked and much more common in males) and this would strongly influence the group result. Any other 'peculiarities' in selection will be clear by this method of recording.

Q2 From the results on grass, plot graphs for each colour of 'insect', of *survivors* against *trial number*: see *figure 156*. You will have to make a second graph for the 'brown dirt' habitat.

Q3 In the 'green grass' habitat, which 'insect' is at a selective advantage?

Q4 How does the 'level of predation' affect this selective advantage?

See *figure 29*.

A green one, because it is better camouflaged from birds (which have good colour vision).

Students may consider that the obvious answer is that as level of predation increases the green insects lose their advantage. This answer seems likely because, as the trials proceed, relatively more green insects are caught. However, the significant point is, *what percentage of each coloured insect survives to breed?*

This information has been calculated and is shown in *table 5*. In the graph in *figure 29*, the percentage of survivors that are green has been plotted against the level

Figure 29
Survivors in the camouflage experiment with red and green toothpicks.

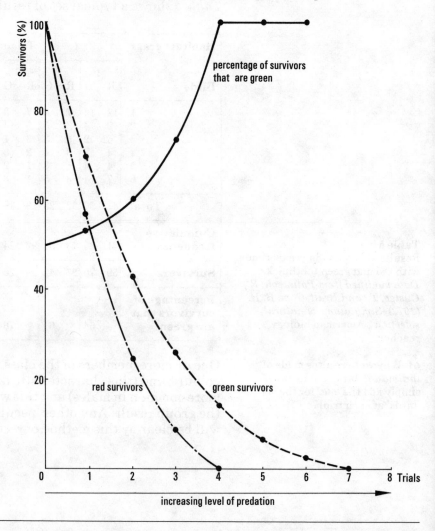

The perpetuation of life

of predation (equivalent to the number of trials). It should now be clear that an increasing level of predation will *tend* to increase the percentage of green insects surviving to breed.

Q5 In the 'brown dirt' habitat which 'insect' is at a selective advantage?

Probably neither red nor green is at any great advantage. In areas where the soil is 'red', there may be a significant selective advantage for red insects.

Q6 Are the results significant (in the numerical sense)? If not, how could you obtain a clearer answer?

This question provides an opportunity to make students evaluate their data critically. They may suggest obtaining more results and this may encourage them to collect the data carefully from other groups in the class.

Q7 What are the major assumptions and limitations of this experiment?

That birds see and 'behave' like people.

Q8 Suppose we accept that the results do imitate the real situation for different varieties of insect. Over a number of generations, what will happen to the insects living in a grassy area?

The proportion of green-coloured insects will increase and the red insects will become less and less common.

Bead model

This experiment investigates the effects of selection over a number of generations.

Using beads with a class
In order to obtain statistically reliable results, at least 400 beads should be used. If time permits, students can do this work either individually or in pairs. However, the procedure can be speeded up by distributing the beads in smaller numbers among the members of the class. The individual results can be combined to form class results which will be statistically reliable.

Thus, in a class of five groups, if each group works with 80 yellow beads and 100 red beads the class total will be 900, which is well above the minimum required.

The results of individual groups may vary considerably from the expected results. However, the class total will approximate closely to what is expected.

The numbers suggested in figure 157 in the *Text* (200 yellow and 250 red beads) are such that 20 per cent selection of yellow insects can be continued for about seven generations without the gene pool falling too far below 400.

Teachers should spend a little time with the class discussing the plan of the experiment (figure 157 in the *Text*) and quick ways of recording and calculating results. It might be possible to weigh beads rather than count them.

Table 6 and *figure 30* show typical results.

| Gen-eration | Yellow insects | | | Red insects | | | | Total Insects |
| | | | | Heterozygous | | Homozygous | | |
	20 % Selected	rr Insects	rr %	Rr Insects	Rr %	RR Insects	RR %	
Parents	–	125	50	–	–	125	50	250
selection *First*	25	100	–	–	–	125	–	225
	–	48	21	104	46	73	33	225
selection *Second*	10	–	–	–	–	–	–	215
	–	43	20	94	44	78	36	215
selection *Third*	9	–	–	–	–	–	–	206
	–	29	14	104	50	73	36	206
selection *Fourth*	6	–	–	–	–	–	–	200
	–	27	$13\frac{1}{2}$	96	48	77	$38\frac{1}{2}$	200
selection *Fifth*	5	–	–	–	–	–	–	195
	–	26	13	88	45	81	42	195
selection *Sixth*	5	–	–	–	–	–	–	190
	1	21	11	88	46	81	43	190
selection *Seventh*	4	–	–	–	–	–	–	186
	–	19	10	84	45	83	45	186
selection *Eighth*	4	–	–	–	–	–	–	182
	–	17	9	80	44	85	47	182

Table 6
Results in the bead model of
selection.

Figure 30
Graph showing frequency of
genotypes against selection in an
experiment to imitate 20 per cent
selection.

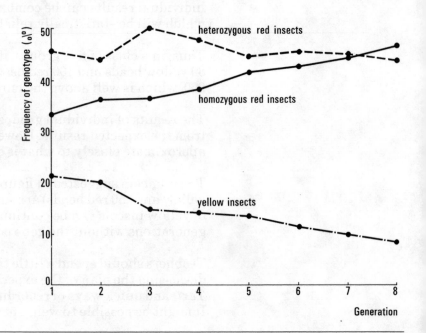

The perpetuation of life

Clever pupils will be able to find fault with the plan in figure 157 in the *Text*. A few simplifications have been made to help pupils with the calculations. Strictly the first generation (starting with 125 homozygous red and 100 yellow) is not achieved by random breeding. The model is still good.

Q9 What changes occur in succeeding generations?

The percentage of yellow insects falls (21 per cent to 9 per cent in eight generations).
The percentage of red (homozygous) insects increases (33 per cent to 47 per cent).
The percentage of red (heterozygous) insects fluctuates but in eight generations shows little trend. With further selection against yellow, the percentage of heterozygous red insects will fall.

Q10 What do you conclude from this result?

After several generations the yellow insects may become extinct. Owing to the presence of heterozygotes this may be at first only 'temporary extinction' and yellow homozygotes will reappear.
Selection has its influence by changing the proportions of the genes in the gene pool. Thus, the proportion of genotypes, and hence phenotypes, is changed in each generation. They do not remain constant.

At this point it may be profitable to discuss the importance of heterozygotes as a pool of recessive genes which may become advantageous if conditions change. The use of the Hardy–Weinberg equation to calculate the proportion of heterozygotes can also be reconsidered (or even introduced) at this stage.

Selection and sickle-cell anaemia

This problem could be tackled in conjunction with section 9.4. The results will show the role of heterozygotes in maintaining 'sterile' homozygotes in a population. It should indicate to students that the characteristics of a future generation are not merely determined by the observable features of their parents. It is the *complete* gene pool of a population that determines the nature of a future generation.

Of course, it is possible that selection against a homozygote will lead to the eventual extinction of a gene. However, the process is often much slower than most students imagine. Indeed it appears to be extremely difficult to get rid of a gene from a population. The problem of eliminating damaging genes, for example, those causing mental deficiency, from a human population is relevant here.

Artificial selection

With only a few bulls being used:
a They might all get some disease.
b If conditions or requirements change, the available 'gene pool' for selection will be small.
c Any harmful or otherwise disadvantageous genes not recognized in a bull will be transmitted to large numbers of the next generation. These will include recessive genes not visible in the phenotype and genes controlling milk yield and quality.

Most mutations produced by radiation will be damaging. Therefore large numbers of organisms would have to be irradiated to yield beneficial mutations. With farm animals this would be too expensive and unethical.

Q3 Summarize the 'message' of Gilbert White's letter in not more than 150 words.

Q4 Is Gilbert White's view of the importance of plants still valid today?

For discussion: this question provides a good chance to revise food chains and to discuss such quotations as 'all flesh is grass'.

Books

*Reading suitable for students

*Biological Sciences Curriculum Study (1971) *High School biology*. 2nd edition. Murray. (Contains a very good section on population genetics.)
*Biological Sciences Curriculum Study (1973) *Molecules to man*. 3rd edition. Edward Arnold. (Contains a very good section on population genetics.)
Brierley, J. K. (Ed.) (1964) *Science in its context*. Heinemann. (Chapter 2, 'Evolution and human biology', contains useful material for class discussion.)
Brierley, J. K. (1964) Modern Science Memoir No. 42. *Some notes on the teaching of evolution in the field, garden, and laboratory*. Murray.
*Dowdeswell, W. H. (1963) *The mechanism of evolution*. Heinemann. (Deals with many aspects of selection.)
Ford, E. B. (1964) *Ecological genetics*. Methuen. (An up-to-date, advanced account of laboratory and field work in population genetics.)
Gabriel, M. L., and Fogel, A. (Eds) (1955) *Great experiments in biology*. Prentice-Hall. (See G. H. Hardy's original paper 'Mendelian proportions in a mixed population', p. 295.)

*Kelly, P. J. (1962) *Evolution and its implications*. Mills & Boon. (Contains relatively elementary accounts of natural and artificial selection.)

Nuffield Secondary Science (1971) *Theme 2 Continuity of life*. Longman. (Contains much material useful to this chapter.)

Open University (1971) Science Foundation Course, Unit 19 *Evolution by natural selection*. Open University Press.

Open University (1971) Science Foundation Course, Unit 20 *Species and population*. Open University Press.

Savage, J. M. (1969) *Evolution*. 2nd edition, Holt, Rinehart & Winston. (A very good summary of modern population genetics.)

Sheppard, P. M. (1967) *Natural selection and heredity*. Hutchinson.

Stern, C. (1960) *Principles of human genetics*. W. H. Freeman. (Contains an advanced treatment of selection in human populations.)

Articles

*Reading suitable for students

Allison, A. C. (1956) 'Human haemoglobin types', *Penguin new biology* series, **21**, 43–58. (Gives a detailed account of the nature of sickle cells and their influence.)

Crosby, J. L. (1961) 'Teaching genetics with an electronic computer'. *Heredity* 16.3, 255–73.

*Demaine, C. (1964) 'Genetics in a school population'. *Biol. Hum. Aff.* **29**, *2*, 29–31 (An account of a survey, undertaken by a grammar school boy, of the frequency of some inherited characteristics.)

Dudley, B. A. C. (1973) 'The mathematical basis of population genetics'. *Journal of biological education* **7**, 41–3.

Dudley, B. A. C. (1972) 'Teaching the Hardy–Weinberg law'. *Journal of biological education* **6**, 359–67.

Jones, R. (1964) 'Some educational aspects of the sickle cell trait'. *Biol. Hum. Aff.* **30**, *1*, 27–31. (An interesting account of a project carried out in an African school; contains useful background material.)

*Kettlewell, H. B. D. (1959) 'Darwin's missing evidence'. *Scientific American* Offprint No. 842. (A fuller treatment of the theme of section 9.2.)

*Lack, D. (1953) 'Darwin's finches'. *Scientific American* Offprint No. 22.

Patterson, R., Custer, T., and Brasstrom, B. H. (1972) 'Simulations of natural selection'. *The American biology teacher* February 1972.

Sheppard, P. M. (1964) 'Protective coloration in some British moths'. *The entomologist* September 1964, **97**, Part 1216, 209–16.

Turner, E. R. A. (1960) 'Survival values of different methods of camouflage as shown in a model population'. *Proc. zool. Soc. Lond.* **136**, 273–84.

Film loops

The following Nuffield O-level Biology loops, published by Longman:
  'Pollination by insects' NBP–2
  'Pollination by wind' NBP–1
  'Selection by predation' NBP–54.
The following Nuffield Secondary Science loops, published by Longman:
  'Artificial insemination of cattle' 0 582 24342 4
  'The breeding of roses' 0 582 24318 1
  'Cereal husbandry' 0 582 24341 6
  'Progeny testing' 0 582 24360 2
  'The results of the selective breeding of two varieties of hen'
  0 582 24312 2
'Predation and protection in the ocean'. BSCS No. 17. John Murray.

Pamphlets

Better breeding (periodical). Milk Marketing Board, Thames Ditton, Surrey.

Outline of the work on evolution
(Chapter 10)

Introduction

10.1	**The voyage of the *Beagle***

10.2	**The deductive method of Darwinism**

Development

10.3	**The evidence for evolution**

10.31	Evidence from geographical distribution

10.32	Evidence from classification and comparative anatomy

10.33	Evidence from fossils

10.34	Evidence from comparative embryology

10.35	Evidence from genetics

10.4	**Evolution of man**

10.41	The classification and comparative anatomy of man

10.42	Changes in the skull bones

10.43	Changes in cranial capacity

10.44	Changes in skeletal features associated with walking

Spine and centre of gravity; pelvis and locomotion; bipedal gait and foot bones

10.45	The evolution of the hand and the use of tools

10
Evolution

Objectives of this chapter

1 to provide the historical
background to Darwin's work on
the problem of evolution and to
illustrate the emphasis placed on
observation and careful deductions
in scientific investigations

2 to provide the student with data
from which he can draw some
conclusions

3 to provide a framework within
which the various theoretical
pieces of evidence for evolution
can be analysed and investigated

4 to use the opportunity to revise
earlier related work (on classification, genetics, and
development) and to integrate it
within the framework of the
theory of evolution

To leave students with the confidence of newly acquired
knowledge and, at the same time, that feeling of
uncertainty which is the stimulus to further study, is one
of the great arts of teaching. It is not easy. Yet, it is hoped
that the form and content of the work of this course will
help teachers to do this.

Ideally, at the end of the course, students should have
acquired three main facilities:
They should understand the concepts embodied in each
chapter (expressed in the *Guide* in the Objectives) and be
able to apply them to fresh problems.
They should understand the methods used in scientific
investigation at the level outlined in the *Text*, and have an
attitude of enquiry that will impel them to carry out further
investigations.
They should recognize the implications of the topics they
have dealt with, both for human society and for an
understanding of living things.

The idea of evolution provides the cornerstone of the whole
edifice of modern biology and it is fitting that it should be
introduced to the students at this stage. Through their
studies they will have gained a knowledge of selection and
genetics that should allow them to understand the nature
of the process of evolution.

Introduction
(Sections 10.1 and 10.2)

10.1 The voyage of the *Beagle*

Q1 What explanation would you
suggest for this diversity of life on
a few small islands in the remote
Pacific?

At this stage students' ideas will probably not be very clear.

10.2 The deductive method of Darwinism

Students could be presented with a model of the growth
of a population. A good example of such work can be found
in investigation 2.1 of Biological Sciences Curriculum
Study, *The world of life: the biosphere*. (See the Reference
material on page 191.) The analogy between the sparrows
in that exercise and the growth of the population of finches
on the Galápagos is clear.

Q1 If a pair of frogs lived for a sexually mature life of 3 years, approximately what percentage of the eggs that they produced would you expect to survive to develop into sexually mature adults in a balanced community?

About 4500 eggs might be laid in three years. Only two need survive to adult life, to replace the parents. Therefore:

$$\frac{1}{2250} \times 100 \text{ per cent} \approx 0.04\%$$

Q2 When would you expect to get an increase in the numbers of a population as indicated by section A–B in *figure 169*?

When a species colonizes a new area or exploits a new food source. If the predators of a population are removed.

Q3 What factors, physical and biotic, might determine the population level at B–C?

Physical factors: available space, temperature, light, oxygen etc.
Biotic factors: available food, quantity of prey and predators, parasites, etc.

Q4 Should the knowledge that you have built up concerning inheritance lead you to renounce the idea that stretching the legs or neck could affect the DNA of the gametes, thereby affecting later generations (see *figure 171*)? Give your reasons.

This question provides an opportunity for students to revise their knowledge of chromosomes and DNA. It is very difficult to find any satisfactory explanation which might support Lamarck's theory.

Development
(Section 10.3 to end of chapter)

10.3 The evidence for evolution

10.31 Evidence from geographical distribution

When, by evolution, a population acquires adaptive characteristics which enable it to exploit an available environment more efficiently than the existing inhabitants, an adaptive radiation of types often occurs. The adaptive radiation of the placental mammals at the beginning of the Cainozoic era is a good example of this process (see figure 175 in the *Text*). A discussion of the reasons for the extinction of the dinosaurs, the dominant vertebrates for more than 100 000 000 years, at the end of the Cretaceous period, could be made at this point.

An analysis of adaptations of mammals which enabled them to fill, so successfully, the niches of the dinosaurs, will probably arise out of a class discussion. The advantages of the placenta and the nourishment of the embryo within the uterus should be made clear. Students should appreciate that there is a relationship between the maturation of the complex mammalian brain and the longer period which a young mammal needs to develop and to become free of parental care.

The adaptive radiation from generalized mammals which lived at the beginning of the Tertiary period gave rise to the various modern orders (see figure 172 in the *Text*). The distinguishing features between these orders could form part of a class discussion. A revision of food habits and the terms 'insectivores', 'carnivores', 'herbivores', and 'ungulates' could prove valuable at this point. A discussion on the modification of body forms, limb structure, and the organization of skulls and teeth in the various orders to their particular ways of life is essential. An example of radiation within an order is given in section 10.33.

A study of the isolation of the Australian continent from the rest of the world and of the ways in which placental and marsupial mammals have evolved in that continent enables us to study the phenomena of divergence and convergence. The Australian mammals which occupy similar habitats to the European mammals show similar features. These adaptations to life in similar habitats are an example of convergence and are illustrated in the Nuffield Secondary Science film loop, 'Australian and British mammals compared' (see the Reference material on page 192).

Darwin's finches

The model mentioned in 10.2 concerning the population growth of house sparrows can be used to explain the evolution of Darwin's finches (Geospizidae) on the Galápagos Islands. Probably soon after these volcanic islands arose from the sea and acquired enough vegetation for birds to survive on them, a small number of generalized finch-like birds arrived from the South American mainland. Since they had no competitors and few predators they multiplied rapidly.

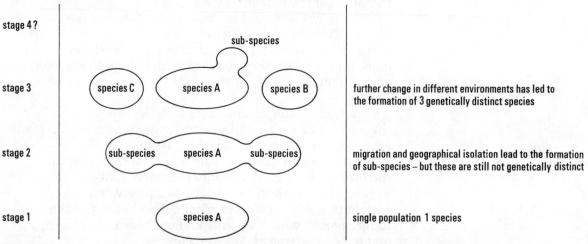

Figure 31
A possible sequence of events leading to the production of different sub-species and species.

As the population expanded on the Galápagos, the descendants of the original finches would make similar journeys to other islands in the group, with different

The perpetuation of life

environmental conditions. The new populations would multiply rapidly in isolation from one another, and by the process of natural selection each would become adapted to its environment.

Some of the populations became adapted to perching in trees and foraging for insect larvae and pupae by removing the bark with their sharp beaks. Others became ground feeders with moderately sized beaks and lived on small seeds and slower moving insects. Large, powerful beaks were evolved by other isolated populations which were capable of crushing tough seeds. (See figure 168 and plates 14 and 15 in the *Text*.)

10.32 Evidence from classification and comparative anatomy

Q1 What are the various kinds of locomotion exhibited by vertebrates?

Students will need to examine carefully the various modifications of the pentadactyl limb, to give an adequate answer based on groupings such as the following:

fish	horse
frogs	cow, sheep
crocodiles, lizards, snakes	bat
pigeon, penguin, ostrich	seal, whale
perching birds	otter
man, and monkey	mole

Figure 32 shows how homologies are used by a taxonomist. The students could be given Step 1 and asked to work out the various groups such as species, genera, families, and orders. Individuals A and B have so many homologous features that they are described as belonging to the same inter-breeding population or species. Eventually, through steps 2 and 3, and finally to step 4, a number of processes can be explained. These include isolation, adaptive radiation, and the general process of evolution.

The evolution game

A valuable and stimulating evolution session could involve the construction of an evolutionary tree on a large sheet of paper using a collection of screws, nails, tacks of various sizes, etc. The following rules are all that the student needs:

The rules of evolution
1 Evolutionary trends usually do not reverse.
2 The trend is from *simple* to *complex*, *and*
3 from *small* to *large*.
4 Usually one characteristic changes at a time.
5 Things like each other are usually more closely related than things unlike each other, but *convergence* is possible.

Figure 32
Steps used in estimating taxonomic
hierarchies and phylogenetic
relationships.
After Hanson, E. D. (1964)
Animal diversity, *Prentice-Hall.*

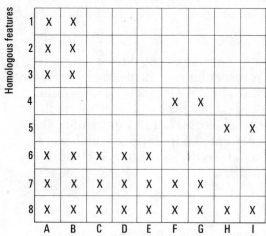

Step 1 *Describe the distribution of homologous
structures among individuals*

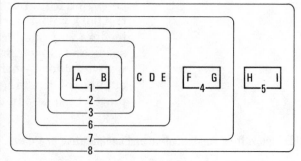

Step 2 *Group individuals on the basis of
homologous structures*

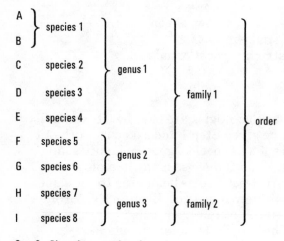

Step 3 *Place the groups in order*

Step 4 Diagram of the phylogenetic relations involved

10.33 Evidence from fossils

Q2 What are the main changes
in the horse's evolution associated
with
a size
b legs and feet
c numbers of toes?

a Increase in size.
b Legs become longer by elongation of the terminal bones
of the foot.
c The animal eventually walks on a single toe in each foot.

Background information and additional work
The work which follows has been adapted from the
Biological Sciences Curriculum Study, *Patterns in the
living world* (see the Reference material on page 190).
Investigation 4.1 in that book, 'Palaeontological
comparisons', illustrates one of the methods used by
palaeontologists for determining the relationships within
a group of organisms.

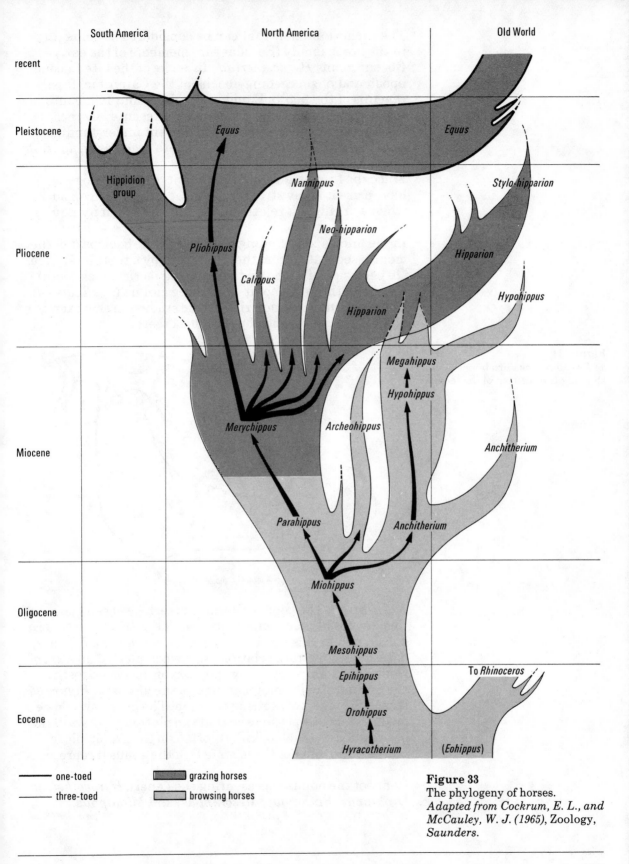

South America   North America   Old World

recent

Pleistocene   *Equus*   *Equus*

                                                   Stylo-
                          Nannippus   *hipparion*

Pliocene   *Neo-hipparion*
            Pliohippus   *Hipparion*
Hippidion
group   *Calippus*   *Hypohippus*

                         Hipparion

                                      Megahippus

                                      Hypohippus

Miocene   *Merychippus*   *Archeohippus*   *Anchitherium*

            Parahippus   *Anchitherium*

                 Miohippus

Oligocene

                 Mesohippus

                 Epihippus   To *Rhinoceros*

Eocene   *Orohippus*

                 Hyracotherium   (*Eohippus*)

—— one-toed   grazing horses
—— three-toed   browsing horses

Figure 33
The phylogeny of horses.
*Adapted from Cockrum, E. L., and
McCauley, W. J. (1965),* Zoology,
Saunders.

The earliest animals that can be considered as belonging to the horse family (Equidae) are members of the early Eocene genus *Hyracotherium*. In rocks of the late Eocene epoch and of succeeding epochs of the Cainozoic, fossil remains of the family Equidae are abundant. Palaeontologists have classified the animals represented by these fossils into about 20 genera. The present understanding of relationships among 17 of these genera is shown in *figure 33*.

Since the fossil material is abundant, palaeontologists have a great many structural characteristics to consider when working out relationships within the family Equidae.

In the horses the grinding teeth are in the back part of the mouth, separated from the front teeth by a toothless space. On each side of each jaw the grinding teeth (cheek teeth) consist of three premolars and three molars (see *figure 34*). The structural characteristic to be studied in this exercise is the *distance* spanned by the cheek teeth.

Figure 34
The skull of a modern horse, showing the position of the teeth.

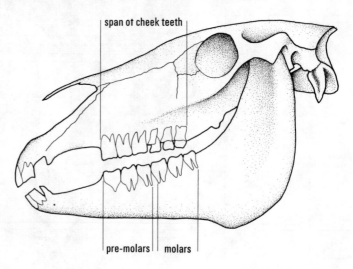

Procedure. The span of the cheek teeth has been measured in many fossil specimens of horses. The data for the genera considered in this exercise are presented in *table 7*. These data suggest certain relationships when plotted on a graph. Construct a graph by using the span of the cheek teeth as the ordinate and geological time as the abscissa. *Figure 35* shows the result of plotting the data. The graph should be made as large as possible so that the plotted points will not be crowded. As each point is plotted on the graph, place beside it the number (from *table 7*) of the genus it represents.

Connect the points representing the genera *Hyracotherium*, *Orohippus*, *Epihippus*, *Mesohippus*, and *Miohippus*.

Table 7

Genera of Equidae	Time of existence	Span of cheek teeth (cm)
1 *Hyracotherium* (*Eohippus*)	early Eocene	4.3
2 *Orohippus*	middle Eocene	4.3
3 *Epihippus*	late Eocene	4.7
4 *Mesohippus*	early Oligocene middle Oligocene	7.2 7.3
5 *Miohippus*	late Oligocene early Miocene	8.4 8.3
6 *Parahippus*	early Miocene	10.0
7 *Archaeohippus*	middle Miocene	11.3
8 *Archaeohippus*	middle Miocene	6.5
9 *Merychippus*	middle Miocene late Miocene	10.2 12.5
10 *Hypohippus*	late Miocene	14.2
11 *Megahippus*	early Pliocene	21.5
12 *Pliohippus*	early Pliocene middle Pliocene	15.5 15.6
13 *Nannippus*	early Pliocene late Pliocene	11.0 10.7
14 *Calippus*	early Pliocene	9.3
15 *Neohipparion*	middle Pliocene	13.1
16 *Hipparion*	middle Pliocene late Pliocene	11.8 11.8
17 *Equus*	late Pliocene Pleistocene	18.8 17.6

Table 7
Span of cheek teeth in the Equidae.

Figure 35
Graph of the measurements of the span of the cheek teeth in the Equidae shown in *table 7*.

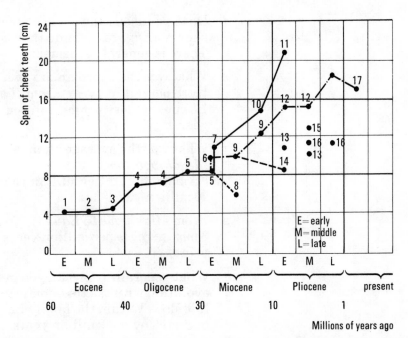

a What seems to have been the trend of evolution in the span of cheek teeth in the Equidae during Eocene and Oligocene times?
To increase from 4.3 cm to 8.4 cm.

b Is it possible to continue a single line farther to the right on the grid?
No.

c Without drawing such a line, describe the trend of evolution in span of cheek teeth during Miocene, Pliocene, and Pleistocene times.
Various trends develop, some increasing in span of cheek teeth rapidly and others not.

Now you need to find out whether the data on span of cheek teeth fit those relationships among the equid genera which have been worked out by palaeontologists from the evidence provided by other structural characteristics.
To do this, draw lines between the dots on your grid to correspond to the arrows on *figure 33*. For example, draw a line from the dot for *Miohippus* to that for *Anchitherium,* continuing through that for *Hyohippus* to that for *Megahippus*; draw another line from the dot for *Miohippus* to that for *Archaeohippus*; draw a third line from *Miohippus* to *Parahippus*, etc.

Conclusions.   If data on any single characteristic conflict with relationships worked out from other characteristics, the data will produce a set of crossing lines when plotted on a graph.

d Do the data on span of cheek teeth support the relationships shown on *figure 33*, or do they conflict with them?
There is general agreement.

e What was the approximate average change in span of cheek teeth per million years from *Hyracotherium* to *Miohippus*?
1.3 mm increase per million years (41 mm in about 30 million years).

f What was the average change from *Miohippus* to *Megahippus*?
6.5 mm increase per million years (123 mm in about 20 million years).

g From *Miohippus* to *Equus*?
3 mm increase per million years (93 mm in about 30 million years).

h What generalization can you make about the rate of evolution within a taxonomic group?
It is slow. Within the family Equidae the changes seen have occurred over 60 million years.

The perpetuation of life

10.34 Evidence from comparative embryology

This has been included as one of the sources of evidence of evolution but is not thought to be appropriate for detailed study at this stage.

10.35 Evidence from genetics

Q3 How does the pattern of inheritance help you to understand the mechanism of evolution?

It enables us to see how chromosome behaviour at mitosis and meiosis can be related to inheritance. The genetic role of fertilization is made clear. We can see how variation is transmitted from generation to generation.

Q4 How does your knowledge of DNA and mutation help you to understand the mechanism of evolution?

The nature of the code for protein synthesis and the way a mutation will be inherited as a change in the DNA all contribute to our understanding. We can recognize how new variations appear. These variations are the raw material for natural selection.

Q5 How would you explain 'the origin of species by means of natural selection'?

This question provides a convenient way of summarizing the work of the whole course. An answer on these lines might be expected.

As a result of the mutation of genes, and through the processes of meiosis and fertilization producing new combinations of genes, a variety of different types of organisms can be produced. Selection acts on this variety of types so that only the better adapted will pass on their genes to succeeding generations. Thus, the composition of a population of offspring will differ to some extent from its parent population.

If populations of organisms become divided into isolated groups it is likely that they will follow different lines of evolution and become distinct from each other. Eventually new types of organisms are provided which are classified as separate species.

The answer could be explained more deeply, of course, and the different roles of the phenotype and genotype and other topics could be taken into consideration.

10.4 Evolution of man

10.41 The classification and comparative anatomy of man

The primates often arouse a keen interest and, in particular, the place of man in this mammalian order can lead to a fresh approach to the importance of classification.

Q1 What are the characteristic features of primates which enable one to place them in the class Mammalia?

They are warm-blooded, with fur or hair, the young are born live (viviparity), the young are suckled on milk from mammary glands, etc.

Q2 How are the legs and arms of man different from those of other mammals?

The line of questioning is intended to show, after careful consideration, that there is no obvious feature which distinguishes primates, including man, from other mammals. This lack of specialization is illustrated by a generalized limb structure which is either adapted for climbing, or has clearly originated from this condition. The digits show general mobility, and both the thumb and toes may be opposable. Primates have flattened nails rather than the hooves and claws of other mammalian orders.

Q3 What do you notice about the relative size and position of the eyes and the general appearance of the front of the face of the primates?

They have eyes at the front of the head. Flattened face.

Q4 In describing the relative importance of the various senses of the primates, emphasize the relationship between the methods of locomotion and feeding, and the 'way of life' in general of the primates.

The large eyes positioned in the front of the head allow for stereoscopic vision which confers the ability to operate well in three-dimensional planes.
The flattened face of the primates in comparison with other mammals leads to a reduction of the organs of smell, and correspondingly there is less importance placed on the sense of smell.

Q5 What significance do you see in the observation that most primates have only one pair of mammary glands?

The fact that most primates have only one pair of mammary glands and that usually only one offspring is born at one time, is a good point for a discussion on the importance of 'parental care' in mammals.

A discussion on the increase in 'parental care' shown in primates over other animals, and the need for such an increased time of dependence on the parents produces some interesting reactions from those who are so dependent: the students.

The progressive development of a larger and more complex brain, especially the cerebral cortex, throughout the primates, should be mentioned. The discussion should be restricted to modern man and apes, as the evolutionary significance of an increase in cranial capacity will be dealt with in 10.43.

All these features can be equated with the modifications necessary for an arboreal life. There is an obvious association that can be made between the reactions necessary for successfully making a way along and between branches, and increased visual capacity, and the development of the brain.

The greater mobility of the digits and the opposability of the thumb are adaptations for gripping the branches, and for picking and manipulating fruit, a major component

The perpetuation of life

of the primate diet. This lack of specialization shown by the hands was of immense importance later in the evolution of man, allowing for the manipulation of tools. The evolution and significance of the hand will be dealt with in 10.45.

10.42 Changes in the skull bones

As an introduction, the significance of certain skull features, particularly those related to the jaws and dentition, can be illustrated by a comparison of a sheep's skull with a carnivore's skull.

At the same time, the position and angle at which the spinal cord is inserted into the cranium through the foramen magnum of a sheep, carnivore, and human could be demonstrated.

It might be advisable to have a class discussion before students are asked to do the questions in order to bring out the significant skull variations.

Pupils are asked to study the 6 skulls in figure 178 of the *Text*.

Q6 List 5 features of skulls which best illustrate the differences between A and some (or all) of B, C, D, E.

In describing the differences in the shapes of the various skulls in this figure, the degree of development of the following features should be emphasized: the size and shape of the forehead, chin, eyebrow ridges, jaws, canine teeth and relative size of face and cranium. (Cranial capacity will be considered in sub-section 10.43.)

Q7 Do the following for each feature you have chosen, denoting the skull it belongs to by letter: write down the order which best represents a sequence from the feature as it is in modern man to the form which looks least like this. The sequence of skulls may not be the same for each feature.

The following is an example of how the observations on one feature would be recorded. It shows how a gradual series between A (modern man), which has the most prominent jaw, and D (*Proconsul*), which has the least prominent jaw, would be built up (for this feature the order of B, E, and C does *not* represent the evolutionary sequence).

Feature of skull: prominence of jaw

    Sequence

Letter	Man
A	Modern man
B	Neanderthal man
E	South African ape-like man – *Australopithecus*
C	Java man (*Homo erectus*)
D	*Proconsul*

The information necessary to tell the age of skulls is given on page 184, but should not be given to the students yet.

The story of the evolution of man can supply interesting examples of the problems of classification, when dealing with the anatomical features of fossils. The degree of variation shown in some fossils can be over-emphasized, as a revision of the intra-specific variation illustrated by the races of man will indicate. The problem of defining a species should be emphasized again as it is crucial in understanding the complexities of man's evolution.

Q8 What are the important differences between the skulls of the gorilla (F) and of modern man (A)?

a Apes have large bony crests for neck and jaw muscles. Man has a lighter skull with no prominent ridges.
b Apes have a heavy lower jaw with large canines and a U-shaped tooth row arrangement (arcade).
Man has a lighter jaw with small canines and rounded arcade.
c Apes tend to have the foramen magnum in a posterior position.
Man has a foramen magnum vertically beneath the cranium.
d Apes tend to have a muzzle.
Man is comparatively flat faced.
e Apes have a comparatively small cranial capacity.
Man has a comparatively large cranial capacity.
f Apes have no bony ridge at the front of the lower jaw; they do not have a chin.
Man has a ridge, that is, a chin.

10.43 Changes in cranial capacity

Measuring the cranial capacity of a skull
We have observed the changes in the skull features of various primates without actually comparing the volume of the space which holds the brain, known as the cranial capacity.

Apparatus and materials
Per class:
skull cut in half – model or human
dry grain or fine sand (approximately 2000 cm^3)
Plasticine
500–1000 cm^3 measuring cylinder
large funnel with a wide spout

Method
a Place the cut half-skull on a flat surface with the open side facing upwards. Put Plasticine underneath the skull in order to keep it steady with the cut surface horizontal.
b Plug any holes with Plasticine, as you are going to fill the cavity with grain.
c Fill the cavity with grain, and check that the grain is level with the top by moving a ruler across.
d Pour the grain carefully from the cavity into the measuring cylinder, using the funnel.
e Record the volume of the half-skull and thereby deduce the cranial capacity of the whole skull.

The sort of figures one would expect for the cranial capacity of an adult human would be between 1000–1800 cm^3, the mean value being about 1300 cm^3.

If it is possible to obtain half sheep's heads, the opportunity could be taken to talk about the structure and functions of the skull and to demonstrate the meaning of the cranial capacity of the skull.

Once the brain has been removed it should be stored in formalin.

Q9 How might you measure the capacity of the cranium if you were provided with a primate skull?

Fill the cranium with some material such as sand or grain (see the method described above) and measure the volume of material required.

Q10 Refer to *figure 178* and, using the grid squares, estimate and record the relative cranial capacity of each of the skulls A–F.

Where man, A = 100
B ≈ 80
C ≈ 75
D ≈ 45
E ≈ 55
gorilla, F ≈ 35

Q11 What deductions can you draw from the data?

Students should be able to make the following relevant comments on the data provided in figures 178 and 179 and table 26 in the *Text*.
a The range of cranial capacities within a species (intra-specific variation) is quite large.
b Hominids have a greater cranial capacity than any of the modern apes.

Q12 By what percentage is modern man's cranial capacity greater than that of *Homo erectus*?

Modern man has a cranial capacity about 30 per cent greater than *Homo erectus*.

Q13 What are these special qualities which distinguish man from the apes?

The sort of comments that will arise from a discussion of the special qualities of man are as follows:
a Power of rational thought, including new power of memory.
b Potential for philosophical thought.
c Consciousness of self.
d Capacity to formulate moral values.
e Capacity for aesthetic pleasure.
f Religious convictions, as shown, for example, in the rites surrounding death.
g Power of speech – and the means of passing on information – giving rise to culture.
h Not only that he uses 'tools', for so do primates, but that he can make them, and do so for future use (see 10.45).
Apes rarely make tools, and then only for immediate needs.

Teachers may prefer to leave discussion of man's special qualities until the end of the chapter.

At this point it is suggested that the teacher should supply more information and give the age of the various skulls, thus linking up sections 10.42 and 10.43. An outline of the evolutionary sequence and the scientific names of man's ancestors could be introduced at an elementary level at this stage.

		Skull	Approximate age of skull (years)
	A	*Homo sapiens sapiens,* modern man	–
	B	*H. sapiens neanderthalensis,* Neanderthal man	70 000
Figure 178 of *Text*	C	*H. erectus,* Pekin and Java man	500 000
	D	*Proconsul*	15 000 000
	E	*Australopithecus* spp.	1 500 000
	F	modern ape	
Figure 179 of *Text*	–	*H. sapiens rhodesiensis,* Rhodesian man	120 000– 40 000
	–	*H. sapiens sapiens* Cro-magnon man (early modern man)	40 000

10.44 Changes in skeletal features associated with walking

In comparing the locomotion and structural features of the apes and man, the following points will arise as a result of discussion.

In apes, the arrangement of thigh muscles and the shape of the hip bones are adaptations for quadruped walking. When they assume a bipedal stance, the hips bend to compensate for the stress in the hip joint. This means that the knees must bend in order to alleviate the resulting forward shift of the centre of gravity. The result is a bent hip and knee gait typical of apes and chimpanzees. The BBC film loop 'Chimpanzee intelligence' (see the Reference material, page 192) might help in enlarging upon the method of movement of the chimpanzee.

For the bipedal gait (typified by man's erect standing gait) to be a successful form of locomotion, the following adaptations are needed to stabilize the body whilst walking.
1 An elongation of the hind limbs with respect to forelimbs.
2 A straightening of the hip and knee joints.
3 A change in the curvature of the lumbar region of the backbone.

The perpetuation of life

4 A modification of the foot, particularly an enlargement of the big toe and its modification to lie along the other toes.
5 A shortening and broadening of the pelvis.
6 A re-arrangement of the muscles of the hips, and particularly an enlargement of the gluteus maximus (the buttocks) which is well developed in man, compared with the apes.

In walking, the hip muscles operate so as to stabilize the pelvis, thus providing a firm base for the leg on the ground. At the same time, there is a small rotation towards the unsupported side which increases the length of the stride. The proportions of the male and female pelvis are different, which means that this rotation is smaller in a woman. One would not be a scientist if one had not noticed that for a given length of stride, a woman is obliged to rotate the pelvis through a greater angle than a man.

The widening of the hips of women at puberty can be explained by the need for a canal wide enough for the birth of a child. However, one has to speculate as to the significance of this hip movement, which is accentuated by the increase in fatty tissue at puberty.

The next questions are on the skeletons shown in figure 180. in the *Text*.

Q14 Compare the positions of the centre of gravity.

The centre of gravity moves progressively further back in early man and modern man. In the quadrupedal ape it is well forward.

Q15 Compare the nature of the spines.

The spine is less curved in early man than in an ape. In modern man the spine is held *vertical* and the curve in the back is further reduced.

Q16 Compare the positions of the knee and hip joints and the attitudes of the legs.

Knee and hip joints become straighter in the bipedal early man and modern man. Legs are held straighter in man.

Q17 Compare the ratios of the length of the hind limb to that of the forelimb.

The ratio is different in the three skeletons. It is approximately:
0.8 in ape (short hind limb)
1.0 in early man
1.2 in modern man (long hind limb)

Q18 What is the function of the pelvis?

To support the hind limbs and join them to the vertebral column. The pelvis must also allow the movement of the hind limbs.

Q19 What do you notice about the place where the hip bone fits into the pelvis, *a* in a man, and *b* in a rabbit?

Q20 What is the significance of your observations in the light of the evidence that apes do not have a 'ball' to the foot?

Q21 In conclusion, using all the information that you have obtained from your observations on skeletal features associated with walking, compare the locomotion of the primates in *figure 180*.

In both, there is a socket to receive the ball-shaped process of the hip bone.

The practical should reveal the importance of the ball and big toes in the striding bipedal gait of man.

Students should be able to relate their observations to the type of locomotion normally seen in, or believed to be typical of, the three primates:

ape   quadrupedal or ungainly, bent, bipedal gait
early man   a forward-bending, bipedal 'jog-trot'
modern man erect, heel to toe, striding bipedal gait

When students discuss the structural adaptations necessary for walking, the following question will probably arise in some form or another: 'What fossil evidence is there which enables one to deduce how these features arose among our hominid ancestors?'

The investigations that follow in this section are included to give students an idea of the type of intensive study that has been carried out on the form of the pelvis, femur, and foot bones of the fossil remains of man's ancestors. This sort of approach has led to judgments about whether the particular hominid was capable of an erect stance or a striding gait.

The transitional stage between an arboreal ancestor and modern bipedal man was probably similar to *Proconsul* (an early chimpanzee-like primate) from East Africa.

Figure 36
The pelvic bones of three primates.
a Modern man.   **b** Ape-like man.
c Ape.
From Le Gros Clark, W. E. (1965), History of the primates, *British Museum (Natural History).*

The shape of the pelvis in the fossils of *Australopithecus* (*figure 36b*) indicate that he probably moved with a bipedal gait, but one that was very different from the heel to toe *striding gait* of *Homo sapiens sapiens*. *Australopithecus* probably covered the ground in quick, short steps with the

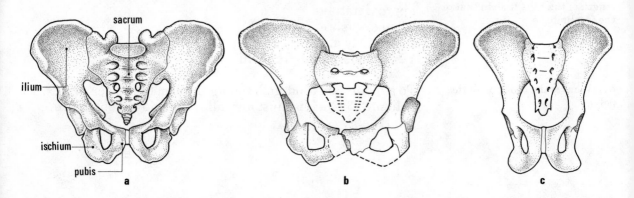

Figure 37
The distribution of mass on the
foot when striding.

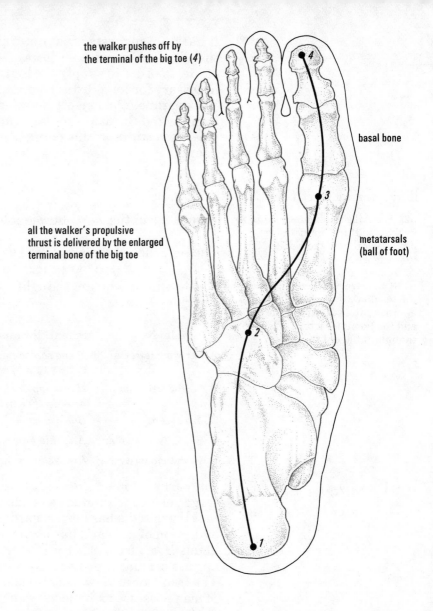

the walker pushes off by
the terminal of the big toe (*4*)

basal bone

all the walker's propulsive
thrust is delivered by the enlarged
terminal bone of the big toe

metatarsals
(ball of foot)

knees and hips slightly bent. The whole effect would be
very similar to a jog-trot, and is an inefficient form of
utilizing the energy for locomotion.

The link between the efficient use of energy and locomotion
and the kind of diet, can lead to a useful revision discussion.

In *Text 2, Living things in action*, the calorific values of a
herbivorous diet and of a carnivorous diet can be analysed.

In the light of the above observations the following question
could lead to a useful discussion. 'How could you explain
the statement that our ancestors probably switched from a
herbivorous diet to a carnivorous diet?'

It has been suggested that natural selection would favour the adoption by our ancestors of a more carnivorous way of life, in order to supply the high energy foodstuffs necessary for long-distance travel.

The possible relationship between the reduction of the canines and the use of tools (weapons) could link up the various sections at this point, although it is a controversial question.

10.45 The evolution of the hand and the use of tools

Q22 Describe the sequence of tools which you think indicates the progression from the earliest type to the more recent.

Q23 Compare your sequence with the teacher's list, which indicates approximately the age of the tools and the hominids which it is thought produced them.

Figure 182 in the *Text* shows tools of different epochs.

The sequence of tools from the earliest to the most modern is D, C, B, A, E (see figure 182).

The hominids who are thought to have produced the tools are given below:

Hominid	Scientific name	Approximate age (years)
A early modern man	*Homo sapiens sapiens* (Cro-magnon)	40 000
B Neanderthal man	*Homo sapiens neanderthalensis*	70 000
C Java man	*Homo erectus*	500 000
D South African man	*Australopithecus*	1 500 000
E recent modern man	*Homo sapiens sapiens*	10 000

The earliest tools were made by striking a few flakes off large pebbles to produce a crude chopping edge. The making of tools has been regarded as a major stage in the evolution of man and has been used to distinguish the genus *Homo* from other primates. The argument is that a certain cranial capacity is essential for this achievement. The traditional view has also maintained that the ancestors of man possessed a hand very similar to that of humans but that it was the limitation of their brains which prevented them from exploiting the hand's potential.

Q24 Compare the sizes and the positions of the thumbs in *figure 184*. (*N.B.* Bones X, Y, and Z in *b* have not been found.)

The fossil remains of a hand similar to *Australopithecus'* show that the thumb is set at a narrower angle and is shorter than that of man.

Q25 What sort of grip do you think that *Australopithecus* possessed?

Although the thumb was opposable, *Australopithecus* probably did not possess a precision grip and used instead a power grip.

Q26 Consider again the brain size of *Australopithecus* (see page 229). Which of the tools in *figure 182* do you think he was capable of making?

With the limitations of shortness and close set of thumb, power grip only (probably), and small brain size, *Australopithecus* could make only the simplest pebble tools (D in figure 182 of the *Text*). It is worth while to note that the apes can only use a power grip and their brain is only slightly smaller than *Australopithecus'*.

There is growing evidence to suggest that the origins of tool making, with their subsequent cultural implications for human evolution, were much earlier and were developed by a less advanced hominid with a less specialized hand than was previously believed.

This argument is presented by J. Napier in 'The Evolution of the hand' (see the Reference material on page 192) and can be summarized by the following quotation:
'The present evidence suggests that the stone implements of early man were as good (or bad) as the hands that made them.'

A discussion on the differences between the brain and hands of the apes and those of modern man, and their significance, will probably bring out the following points:

1 Apes rarely make tools and then only for immediate needs.
Man makes tools to meet future use.
2 Apes do not use fire.
Man uses fire.
3 Apes do not construct large or permanent buildings.
Man builds houses or lives in tents, caves, and so on.
4 Apes show no settled agriculture; they are collectors or hunters.
Man has developed agriculture to meet his need for food (except in the case of some primitive types).
5 Apes do not display cultures.
Man shows various types of culture which involve activities such as speech, art, religion, and science.

Figure 38
A diagram showing possible relationships between the different types of hominoids.

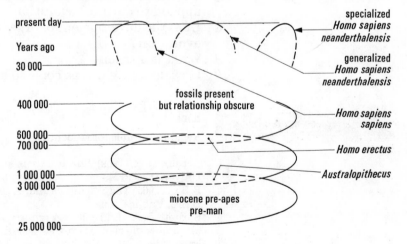

Summary

A useful way of considering the evidence for man's evolution is to compare modern man with modern apes, notice the differences, and then try to trace in the fossil records when these differences arose.

Essay and discussion topics

Most of these will require additional reading to be undertaken. The questions in the *Text* could also serve as topics for this purpose.

1 Why was Darwin's theory of evolution eventually accepted by so many people, despite the hostility it received at first?
2 Both Charles Darwin and A. R. Wallace went on long journeys of exploration. In what way do you consider this to have been significant to them in working out their theories of evolution?
3 Should it make any difference to people's attitudes in everyday life if they accept that man has an ape-like ancestor?
4 It is not possible to demonstrate all the lines of evolution in the past from fossils that have been found. Does this make the theory of evolution invalid?
5 How did it come about that the researches of Darwin, Mendel, and the cytologists were able to proceed independently during the nineteenth century without their inter-relationship being understood?
6 Cultural evolution is a much more rapid process than the evolution of man's physical nature. What effect may this have on the future evolution of man?
7 To what extent can the concept of natural selection be applied to cultural evolution?
8 Does the development of organisms play any part in their evolution?
9 What is a species?
10 How do living things come to be both similar and different?

Reference material

Books

*Reading suitable for students

Arthur, D. R. (1969) *Survival: man and his environment.* English Universities Press.
*B.B.C. (1965) The Science of Man 4 *Heredity and evolution.* B.B.C. Publications.
*Barnett, S. A. (Ed.) (1962) *A century of Darwin.* Heinemann. (A number of scientists review Darwin's ideas in the light of today's knowledge.)
de Beer, G. (1958) *A handbook of evolution.* 3rd edition. British Museum (Natural History).
Biological Sciences Curriculum Study (1972) 'Biology: an environmental

approach'. *Patterns in the living world*. John Murray. (See investigation 4.1, 'Palaeontological comparisons'.)

Biological Sciences Curriculum Study (1972) 'Biology: an environmental approach'. *The world of life: the biosphere*. John Murray. (See investigation 2.1, 'Population growth: a model'.)

Campbell, B. G. (1967) *Human evolution*. Heinemann.

Clarke, R. A. (1971) *Biology by inquiry* Book 3. Heinemann.

Crellin, J. K. (1968) Jackdaw Series No. 85 *Darwin and evolution: a collection of contemporary documents*. Cape. (A useful teaching aid. The various items in this folder could perform a valuable function in the development of this theme.)

*Dale, A. (1953) *An introduction to social biology*. 3rd edition. Heinemann. (Very readable. Contains useful information about human evolution.)

*Darwin, C. *The origin of species by means of natural selection*. (Several editions have been published. It was originally published by John Murray in 1859.)

*Darwin, C. *The voyage of the* Beagle. (There are several editions of this book. It was originally called *A naturalist's voyage around the world* and published by John Murray in 1845. As with *The origin of species*, students can gain a closer insight into Darwin's ideas by reading his original publications.)

*Darwin, C. (1952) *The next million years*. Hart-Davis. (Some interesting comments on possible future evolution by one of Darwin's grandsons.)

*Eiseley, L. (1959) *Darwin's century – evolution and the men who discovered it*. Gollancz.

*Gabriel, M. L., and Fogel, S. (Eds) (1955) *Great experiments in biology*. Prentice-Hall. (Contains some of the classic papers on the study of genetics and evolution.)

*Himmelfarb, G. (1959) *Darwin and the Darwinian revolution*. Chatto & Windus. (Like Eiseley's work, this book provides an interesting account of the social impact of Darwin's theory.)

Howell, F. C. (1967) Life Nature Library *Early man*. Time–Life International.

Howells, W. (1967) *Mankind in the making: the story of human evolution*. Penguin.

*Huxley, J., and Kettlewell, H. B. D. (1965) *Darwin and his world*. Thames & Hudson. (An account of Darwin's life.)

*Kelly, P. J. (1962) *Evolution and its implications*. Mills & Boon. (Contains an elementary account of the social context of Darwin's work.)

*Medawar, P. B. (1960 *The future of man*. Methuen.

*Moore, R. E. (1962) Life Nature Library *Evolution*. Time–Life International. (A 'popular' account of the subject.)

Nuffield Secondary Science (1971) *Theme 2 Continuity of life*. Longman. (See pages 266–7.)

Simpson, G. G. (1951) *Horses*. Oxford University Press.

Stebbins, G. L. (1966) *Processes of organic evolution*. Prentice-Hall.

*Tattersall, I. (1970) *Man's ancestors: an introduction to primate and human evolution*. John Murray.

Articles

*Reading suitable for students

Bennett, A. J. (1964) 'Mendel's laws'. *School science review* **46**, *158*, 35–42. (An interesting comment on the relevance of Mendel's paper.)

*Eiseley, L. C. (1956) 'Charles Darwin'. *Scientific American* Offprint No. 108.

*Lack, D. (1953) 'Darwin's finches'. *Scientific American* Offprint No. 22. (An account of the finches of the Galápagos Islands.)

Napier, J. (1967) 'The antiquity of human walking'. *Scientific American* Offprint No. 1070.

Napier, J. (1962) 'The evolution of the hand'. *Scientific American* Offprint No. 140.

Film

'The voyage of the *Beagle*'. Sound, black and white, 20 minutes. B.B.C. *To hire* B.B.C. Enterprises Film Hire, 25 The Burroughs, Hendon, London NW4; *to purchase* B.B.C. Enterprises, 310 Villiers House, The Broadway, London W5 2PA.

Film loops

'Chimpanzee intelligence'. PA 206. B.B.C. Publications, 35 Marylebone High St, London W1M 4AA.

The following Nuffield Secondary Science loops, published by Longman:

'Australian and British mammals compared' 0 582 24343 2

'Formation of fossils' 0 582 24303 3

'Geological time scale 1' 0 582 24300 9

'Geological time scale 2' 0 582 24301 7

'Motor skills in primates' 0 582 24311 4

The perpetuation of life

Appendix 1
List of chemical recipes

Acetic alcohol

25 cm^3 acetic acid (glacial)
75 cm^3 ethanol or industrial alcohol

Feulgen stain (Schiff's reagent)

200 cm^3 distilled water
1 g fuchsin basic
30 cm^3 hydrochloric acid (molar)
3 g potassium metabisulphite ($K_2S_2O_5$)
Bring distilled water to the boil and add the fuchsin. Shake
well and cool to 50 °C. Add the 1M hydrochloric acid and the
potassium metabisulphite. Allow to bleach for 24 hours in
a tightly stoppered bottle *in the dark*. Add 0.5 g decolorizing
charcoal. Shake thoroughly and filter rapidly through
coarse filter paper. Store in a tightly stoppered bottle in a
cool, dark place.
(This stain can be purchased already prepared.)

Use
1 Fix the material (plant or animal) in acetic alcohol for at
least one hour, preferably one day.
2 Hydrolyse the material by placing it for 6 minutes in
1M HCl already at 60 °C in an oven or water bath.
3 Transfer from the HCl at 60 °C to Feulgen reagent at room
temperature. Leave for one to two hours and chromosomes
and nuclei will stain purple.

Holtfreter's medium

Dissolve:
3.5 g sodium chloride
0.2 g sodium bicarbonate
0.05 g potassium chloride
0.1 g calcium chloride
in 1 dm^3 distilled water

Indole acetic acid

Dissolve 0.1 g of indole–2–acetic acid (IAA) in 2 cm^3 of
ethanol and add this to 900 cm^3 of distilled water. Warm
to 80 °C for five minutes and make up the volume to 1 dm^3.
The solution contains 100 parts per million (p.p.m.). This
stock may be stored in the refrigerator for up to two weeks.
Dilute the stock solution with distilled water to prepare the
experimental solutions.

Media

For Bacillus subtilis

a *Nutrient broth*
10 g beef extract
10 g peptone
5 g sodium chloride
Make up to 1000 cm^3 with distilled water, adjust pH to 7.5
by adding sodium bicarbonate, then filter, and autoclave.
However, the simplest method is to use Oxoid Nutrient
Broth No. 2 (CM 68) tablets. Dissolve one tablet per 10 cm^3
distilled water and sterilize by autoclaving at
102 583 N m^{-2} (15 lbf in^{-2}) for 15 minutes.

b *Nutrient agar*
Add 15 g dry agar to 1 dm^3 nutrient broth and soak for
15 minutes. Heat at 100 °C until dissolved, then filter.
Autoclave.
N.B. *Oxoid nutrient agar tablets* (CM 4) are made to the
same formula. Use one tablet per 5 cm^3 distilled water.
Again soak for 15 minutes and then autoclave.

It is best to make bulk supplies of medium using thick glass
bottles with rubber sealed screw tops, such as medicine flats
(available from Hospital and Laboratory Supplies,
12 Charterhouse Square, London EC1).
Before putting these in the autoclave, loosen the screw
tops about a quarter turn. Tighten them as soon as you
take them out. The sterilized medium will keep indefinitely
in these bottles. When plates are to be poured, put the
bottles in a pan of water and bring them to the boil. Cool
them. They can now be kept in a water bath at 45–50 °C
for 1$\frac{1}{2}$ hours. Then nutrient agar can be dispensed in 15 cm^3
portions into sterile boiling tubes plugged with cottonwool.

For Drosophila
See Appendix 2, page 200.

Methylene blue

Mix:
1 g methylene blue
0.6 g sodium chloride
100 cm^3 distilled water

Orcein

This may be dissolved in either acetic acid (acetic–orcein)
or propionic acid (propionic–orcein). For many purposes
propionic–orcein is superior as it tends to stain the
cytoplasm less than acetic–orcein.
Place 45 cm^3 glacial acetic (or propionic) acid with 55 cm^3
distilled water in a flask, preferably fitted with a reflux

condenser. If it is not, put in a fume cupboard. Add 1–2 g synthetic orcein and some boiling stones or glass beads. Boil gently for 30–60 minutes (if fitted with a reflux condenser, for 10–15 minutes). Allow to cool and filter. Prepared acetic–orcein can be obtained from biological supply agencies.

PTC papers

Dissolve 1.3 g of PTC (= phenylthiocarbamide = phenyl-thiourea) in boiling water and make up to 1 dm^3.
Soak a filter paper in the solution. Remove and dry it. Cut it into strips 1 cm × 2 cm. The strips can be stored indefinitely in an envelope or covered dish.

Saline, for insects

0.75 % sodium chloride
Dissolve 7.5 g sodium chloride in distilled water and make up to 1000 cm^3.

Appendix 2
Animals and plants

Experiments and observations using live animals
The 'Act to amend the law relating to cruelty to animals' (1876) regulates and restricts experiments on living *vertebrate* animals, that are calculated to cause pain. There are no statutory definitions of 'experiments', 'living', 'calculated', or 'pain', but it has been established and published policy for many years to regard the Act as applying wherever it is proposed to subject an animal to interference that carries any risk, however remote, of disturbance to its normal health and comfort, by a procedure which is not intended to be therapeutic, and the outcome of which is uncertain or unknown.

It follows from this liberal interpretation that relatively *painless* experiments have to satisfy the primary restrictions in the Act, that they must be performed either 'with a view to the advancement, by new discovery, of physiological knowledge, which will be useful for saving or prolonging life, or alleviating suffering' – section 3(1) – or be certified to be 'absolutely necessary for the due instruction' of the person to whom they are demonstrated, 'with a view to their acquiring physiological knowledge or knowledge which will be useful to them for saving or prolonging life, or alleviating suffering' – section 3, proviso (1). The Act sets a high standard of justification (and control) for undertaking experiments, many of which are in the result, and may often be in prospect, seemingly painless.

It also has to be borne in mind that it is an offence under the Protection of Animals Act (1911), to cause unnecessary suffering to an animal. There is an exemption for experiments licensed under the Cruelty to Animals Act (1876), but not for any other experiments.

The experiments described in this course do not contravene the Acts mentioned above. Thus, for example, the one that involves feeding tadpoles with thyroxin is not an experiment in the meaning of the Acts, because the result is predictable. Rearing tadpoles or other animals in different temperatures, again does not contravene the Acts, provided the temperatures are not extreme. Humane killing resulting in a swift death does not contravene the Acts.

The Acts themselves are not specific enough to allow a judgment to be made on particular experiments, and if there is the slightest doubt about the legality of a proposed experiment, it is well to consult the Home Office.

Besides the problem of legality, there is the question of ethical standards in dealing with live animals. Invertebrate animals are not covered by the Acts, but clearly if we wish our students to develop a respect for life, a humane approach should be applied to these as well as to vertebrates. Frequently the crucial educational problem is not the nature of an observation or experiment – which may be harmless enough – but the students' attitude towards the animals during the operations. It is part of the purpose of a course which makes frequent use of live animals, to instil in the students a balanced, humane approach to the wellbeing of the creatures they study.

Bryant, J. J. (1967) *Biology teaching in schools involving experiment or demonstration with animals or with pupils.* The Association for Science Education.
Lane-Petter, W. (Ed.) (1963) *Animals for research. Principles of breeding and management.* Academic Press.
Universities' Federation for Animal Welfare (1967 *Handbook on the care and management of laboratory animals.*
Wray, J. (1973) *A recommended practice for schools relating to the use of living organisms and material of living origin.* English Universities Press for the Schools Council.

Amphibia

Induced breeding

The tadpoles of the common frog (*Rana temporaria*) are only obtainable in the wild during the late spring or early summer. Sometimes during the winter months adult specimens can be induced to breed by pituitary injection. However, this is an uncertain procedure. Edible frogs (*Rana esculenta*) can usually be obtained from suppliers through most of the year, and these can also be induced to breed sometimes during the winter. *Rana pipiens* is probably the best *frog* to use for induced breeding during the winter months. *Xenopus laevis*, the African clawed toad, can be easily bred from at all times of the year, provided that gonadotrophic hormone is injected. Both the adults and tadpoles are easy to keep. *Xenopus* is probably the most suitable amphibian for laboratory work. See 'Xenopus laevis' in Appendix 3 of *Teachers' guide* 1, *Introducing living things,* and page 62 of Nuffield Advanced Biological Science *Laboratory book.*

The method for the induction of breeding, with *Xenopus*, is to inject a form of chorionic gonadotrophin. This can be purchased in ampoules and is marketed under various proprietary trade marks. Chorulon and Pregnyl are two kinds, sold in ampoules for injecting into toads. For doses sufficient to produce amplexus and subsequent egg-laying and fertilization see *table 8*. The hormone should be injected into the dorsal lymph sac.

Xenopus toads should not be fed within a day of mating, or they tend to be sick. It is best to place a non-metallic platform in the mating tank about 5 cm from the bottom with a clear space about 6 mm around the sides. This allows the eggs to fall beneath the platform where they cannot be eaten.

Table 8
Injection programme for inducing ovulation and fertilization in *Xenopus*.

Time	Ampoules to be used	Dose for male	Dose for female
3–4 days before ovulation is required	2 × 100 i.u.	50 i.u.	100 i.u.
1 day before ovulation	500 i.u.	150 i.u.	350 i.u.

A possible timetable that can be used with *Xenopus* is to give the toads the second injection in the morning, place them in the mating tank at about 16.00 hours, and collect the eggs next morning.

Figure 39
A mating tank for *Xenopus*.

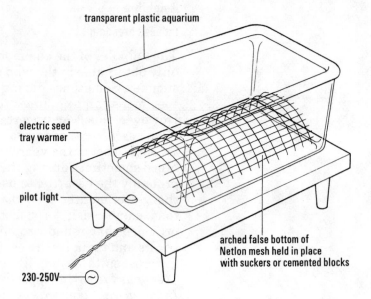

transparent plastic aquarium

electric seed tray warmer

pilot light

230-250V

arched false bottom of Netlon mesh held in place with suckers or cemented blocks

The perpetuation of life

Rearing eggs and tadpoles

Clean pond water known to be used by frogs for breeding in naturally is the most suitable medium in which to keep most species of Amphibia in the laboratory.

Tap water can be used but may need to be filtered through charcoal and sand or allowed to stand for a few days. This removes chlorine, which is toxic to amphibian larvae.

Holtfreter's solution (see Appendix 1, page 193) is also suitable.

Water that has run through copper pipes is toxic to tadpoles. It is best to keep eggs and tadpoles in glass containers, which should be previously sterilized or washed in hot, soapy water and then rinsed in clean running water for two hours or more.

For eggs of all species, and the tadpoles of frogs, fill the containers to a depth of 2.5 cm with the culture water. *Xenopus* tadpoles can be kept in greater depths.

For most species the water has to be changed every three or four days but this is unnecessary with *Xenopus* because of the low oxygen requirements of its tadpoles.

Most species develop well at about 20 °C. Allow the water to reach this temperature before placing eggs or tadpoles into it.

Do not crowd eggs or larvae. Overcrowding before hatching produces abnormalities, and with tadpoles it leads to contamination and poisoning. Two eggs or tadpoles should be allowed approximately 40 cm^3 of water.

Until the tadpoles' mouths open, no food is necessary. Frog tadpoles are herbivorous and can be fed on slightly boiled lettuce or spinach. *Xenopus* tadpoles thrive well on dried nettle powder sprinkled on the water.

Metamorphosis occurs in most species around the eleventh week under laboratory conditions, depending on the species. In *Xenopus* it may occur as early as the sixth week.

At metamorphosis, some means by which the young frog can crawl out of the water must be provided. As adults, frogs are best kept in vivaria containing a large shallow dish of water. *Xenopus* adults are aquatic and, providing the water is shallow enough (about 15 cm) for them to swim to the surface, they can be kept in aquaria.

Most adult frogs require a good supply of live insects or worms for food. They always feed on land. *Xenopus* adults prefer live meat such as small fish, tadpoles, worms, etc. but feed well on small pieces of liver or minced meat dropped into the water.

Take care to remove uneaten food after an hour or so to prevent contamination.

Xenopus adults can escape from containers if the water is shallow enough to permit jumping or if they can crawl up the sides. A few hours out of water are fatal. For this reason, it is a sensible precaution to have firmly fixed wire-netting tops on aquaria.

Leadley Brown, A. (1970) *The African clawed toad, Xenopus laevis: A guide for laboratory practical work.* Butterworth.
Nuffield Biology (1974) Teachers' Guide *Introducing living things.* Revised edition. Longman.

Drosophila, culture of

The best arrangement for saving time and labour and yet keeping the cost of materials to a reasonable level is to make the culture medium at school but to buy separate stocks of males and virgin females as required. Sexing the young flies is easily the most time-consuming and inconvenient aspect of culturing *Drosophila* for school use.

Pure-line *Drosophila* cultures and crosses already set up can be obtained from suppliers. These should arrive in small tubes containing a small amount of culture medium and viable larvae and pupae. It may be best to ask for the adults to be removed before transportation. If they are sent they are liable to get stuck and possibly crushed. Further, if it is a cross, the flies will have to be removed shortly after they arrive to prevent confusion with the offspring.

Other strains of *Drosophila* that might be used instead of the ebony body strain for this work are: Vestigial wing (vg) – very small wings, but they are less viable and tend to give poor ratios in F_2. Scarlet (st) – with bright orange-red eyes that darken as the fly gets older.
Dumpy (dp) – wings two-thirds of normal length and appearing as if the ends had been cut off.

The perpetuation of life

1   *Preparing the culture medium*
There are several alternative recipes but the following is reliable and effective.

Mix 100 g of maize meal, 30 g of powdered agar, 26 g of dried yeast, and 50 g of brown sugar with 1600 cm^3 of water. Stir and gently boil until a uniform consistency is obtained.

Stir 0.5 g of nipagin in approximately 80 cm^3 of water and add it to the medium just before pouring it into the bottles. The quantity will fill about 20 bottles (200 cm^3) to a depth of 2.5 cm.

Nipagin is included to inhibit fungal growth. It is methyl hydroxy benzoate (the methyl and propyl forms are both suitable). 2 or 3 cm^3 of 0.5 per cent propionic acid in 500 cm^3 of culture medium is an adequate substitute.

Culture media for *Drosophila* can be obtained already prepared and sterilized in culture bottles. Culture media, if first sterilized, can be stored in a refrigerator for about six weeks. In a deep-freeze they can be stored for six months.

Instant *Drosophila* medium is supplied by T. Gerrard & Co., Gerrard House, Worthing Road, East Preston, Littlehampton, Sussex. This requires *no heating or sterilizing*. It is poured into sterile culture tubes in dry form; then water and a pinch of yeast (provided) are added. This is dearer but may be more convenient in some cases.

2   *Preparing stock bottles*
Small bottles (approximately 200 cm^3) or large flat-bottomed specimen tubes 13 cm × 4 cm diameter or larger) can be used. They should be thoroughly washed with hot water containing a combined disinfectant and detergent, used as directed and rinsed with cold water. Care should be taken with hot disinfectants, which can be caustic.

Figure 40
A *Drosophila* culture bottle set up. The adults have been removed and a new generation will hatch from the pupae.

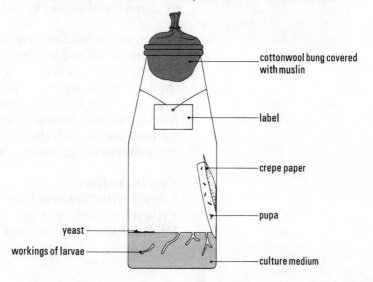

cottonwool bung covered with muslin

label

crepe paper

pupa

yeast

workings of larvae

culture medium

The culture medium is poured into the stock bottles while still hot and fluid. With a clean glass rod push one end of a double strip of paper towel, absorbent toilet paper, crepe paper, medicinal gauze, or paper handkerchief to the bottom of the medium while it is still soft. Stopper the bottles with cottonwool covered with muslin tied at the top.

3 *Sterilization*
It is advisable at this stage to sterilize the culture bottles containing medium by keeping them for 15 minutes at $102\,583\ \mathrm{N\ m^{-2}}$ ($15\ \mathrm{lbf\ in^{-2}}$) pressure in a pressure cooker or autoclave.
An alternative but less satisfactory method is to immerse the bottles, after cleaning, in boiling water, perhaps in a water bath, for at least 30 minutes. Another, to prevent breakage, is to heat them gradually to about 160 °C and then keep them at that temperature in an oven for 1 or 2 hours. The culture medium should be boiled for about the same length of time (an extra 20 cm^3 or so of water should be added to allow for evaporation) and precautions should be taken to prevent contamination of other materials used. After sterilization, allow the medium to cool and then add three drops of live yeast suspension to the medium in each bottle.

4 *Transfer of flies*
Before adding the flies, dry off any moisture on the inside of culture bottles with a piece of sterilized absorbent paper. It is probably best to use 5–10 pairs to start a stock culture. However, a mixture of 3 females and 6 males should produce 200 to 300 offspring. A high proportion of males ensures that the females will be mated more or less simultaneously so that the offspring may hatch out at about the same time. By allowing for 80 offspring per female in a stock culture, one can be certain of obtaining the number of flies required for classwork, and, possibly, some extra.

Anaesthetize flies from original stock, sex them if you have not obtained separate stocks (see *Text*, section 2.3 and page 204 of this *Guide*), and place selected specimens on the absorbent paper in the bottle; otherwise they will tend to get stuck to damp patches on the glass or culture medium. Replace bungs immediately. Afterwards a check should be made to see that all the flies have recovered, or to compensate for those that have not. Remove dead flies.

5 *Keeping cultures*
Stock bottles are best kept on trays holding 6–12 bottles each, when they are easily stored, inspected, and handled. Each culture bottle should be labelled carefully.

Figure 41
A label for a *Drosophila* culture
bottle.

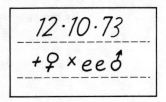

Drosophila breed best at 25 °C. They become sterile if kept above 28 °C for any length of time, and the rate of reproduction is significantly reduced if the temperature drops below 15 °C. Therefore it is best to keep them in an incubator. Normal laboratory temperature is, however, usually adequate, but care must be taken to avoid excessive heat from radiators or the sun, and possible sudden drops of temperature at night. Do not put the culture bottles on window sills. Rapid changes in temperature cause condensation in the culture bottle to which flies get stuck and die.

A makeshift incubator can be made by keeping the bottles in an aquarium or box and insulating them with cotton-wool wrapped round the bottoms of the bottles. An electric light bulb covered with a shade can be used as a source of heat. It is advisable to keep a thermometer by the cultures. Stocks will need to be placed in new stock culture bottles once every three or four weeks. If the culture medium appears to dry out before this time, a few drops of yeast suspension should be added.

Cultures in which moulds start to develop should be discarded immediately. Be careful not to transfer flies from mouldy to healthy cultures.

Breeding cultures for classwork

For class breeding experiments, it is best to use flat-bottomed specimen tubes (2.5 cm × 10 cm). Each can hold 100 flies, although 50 is a better number. Culture medium should be added to a standardized depth of about 2 cm. Otherwise the procedures are the same as those for preparing stock cultures.

At 25 °C the imago hatches from the pupa approximately nine days after the egg is laid. Females can lay eggs 12 hours after emergence (sometimes after 48 hours if etherized). Males are potent about three hours after hatching. The average length of adult life is 26 days for females and 33 days for males.

Table 9 is a guide to the development of *Drosophila melanogaster* at 25 °C.

By hours	By days (approximately)	Stage
0	0	egg laid
0–22	0–1	embryo
22	1	hatching from egg (first instar)
47	2	first moult (second instar)
70	3	second moult (third instar)
118	5	formation of puparium
122	5	'prepupal' moult (fourth instar)
130	$5\frac{1}{2}$	pupa; eversion of head, wings, and legs
167	7	pigmentation of pupal eyes
214	9	adults emerge from puparium
215	9	wings unfold to adult size

Table 9
After Strickberger, M. W. (1962) Experiments in genetics with Drosophila, Wiley.

The life-cycle of *Drosophila* at 25 °C is ten days. It is best, however, to plan classwork on the basis of a 12–14-day life-cycle and to allow sufficient time in between each life-cycle for sexing and the transfer of flies.

In order to ensure that the requirements for classwork are met in time, breeding can be staggered. For example, when the offspring of a cross are needed, it is best to set up separate batches of cultures 12 or 14 days and 9 days previously.

Reciprocal crosses in separate culture bottles should be used, ee ♀ × + ♂ in some culture bottles and + ♀ × ee ♂ in others. This standard procedure also helps to overcome preconceptions about the influence of sex on inheritance.

Female *Drosophila* retain sperms in sperm sacs and thus virgins must be used when setting up crosses. Virgin males, of course, are not necessary.

Obtaining virgins

To obtain virgin females, all the adults are removed from a culture. Any flies emerging within the next 9 hours are likely to be virgins. However, for up to 12 hours after hatching most females remain virgin. Under school conditions a suitable and reliable procedure is to remove the adults first thing in the morning (say 9.00 hours) and to harvest the virgin stock in the afternoon (preferably before 17.00 hours). An additional culling at lunchtime is ideal. The procedure should be repeated on successive days, or when it is convenient, until all of the culture has hatched. Virgins should be sexed immediately, and stocks of virgin females kept for use at an appropriate time. (If time is scarce they can be purchased from biological supply agencies.)

Young virgin females have a pale, almost white, body colour. They tend to hatch earlier than the males. When in doubt, confirm the identification of a female by checking that the sex comb is absent. Other criteria for sexing are the brown chitinized claspers at the end of the abdomen in the male, which are absent in the female.

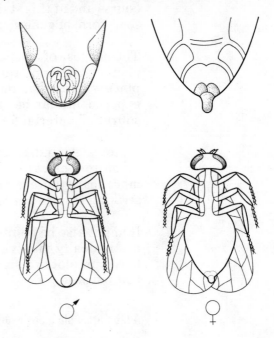

Figure 42
Distinguishing male and female *Drosophila*. In recently emerged adults there is little pigmentation in the abdomen and so the chitinized claspers of the male stand out clearly. They are absent in the female.
After B. J. F. Haller.

Handling *Drosophila*

The methods and apparatus used for handling *Drosophila* are dealt with in the *Text*. Glass (rather than plastic) filter funnels and specimen tubes are recommended as they allow the flies to be seen clearly. Very few are broken by students.

Specimen tubes of 1.5 cm in diameter, and filter funnels with 5 cm diameter mouth are suitable sizes for the etherizer. After a little experience it is easy to judge the most suitable amount of ether–ethanol to be used. The cottonwool should be lightly moist with ether–ethanol but not wet. If in doubt, start with one or two drops from a pipette and gradually increase the amount, if necessary, until it is effective.

Directly the flies are immobile they should be removed from the etherizer. The flies will usually remain etherized for about seven minutes. Dead flies, or those to be discarded after counting, can be put into a 'morgue' – a spare bottle containing 2 or 3 cm of 70 per cent alcohol. They can then be used for demonstrating characteristics and other purposes. If the flies are not required any more, 2 or 3 cm of motor oil in the bottle will do. A paint brush can be used for picking up dead flies.

Haskell, G. (1961) *Practical heredity with* Drosophila. Oliver & Boyd.

Garden and greenhouse

All the plants mentioned as suitable for work in this course can be kept in the school garden, greenhouse, or laboratory. Plants will play a large part in the work of the course and at this stage one should consider setting up some form of genetic garden.

The purpose of a genetic garden is threefold: to provide plants for demonstrating topics in the course; to provide a place where, sometimes over a long term, breeding experiments can be carried out; and to provide a ready source of material for cytological work in the laboratory.

A plot of ground, an area in a greenhouse, or a bench in a laboratory can all serve as a site for a genetic garden. The more space there is, the more extensive it can be, but striking results can be obtained with limited resources.

It might also be pointed out that a genetic garden can be aesthetically satisfying, and there is no reason why a border or plot in a school's garden should not be used for this purpose.

A laboratory plant enclosure

Whatever other facilities are available, it is necessary to have some space in the laboratory where plants can be kept. At the least, this can be used to put pots on, but it can be quite easily transformed into a miniature greenhouse.

A 10–12 cm high rim of wood around an area on a bench will keep in place a layer of gravel. The bench must be sturdy, for the gravel is heavy. The surface of the bench should be made waterproof with a layer of asbestos or other material or, better still, the rim and bench surface can be lined with polythene or other impervious plastic material.

A cover of transparent plastic stretched over a wooden or metal frame will provide protection from dust, and retain moisture and, to a certain extent, heat.

If need be, the enclosure can be heated either by an electric light bulb or insulated and waterproof heating wire laid beneath the gravel. The addition of a thermostat will allow the heating to be regulated easily.

Artificial lighting is sometimes necessary. For most purposes 80 W fluorescent tubes in either the warm white or 3500 K colour will be suitable. Three or four such tubes, depending on the lighting of the surroundings, will give sufficient light for an area 1 m by 2 m.

The perpetuation of life

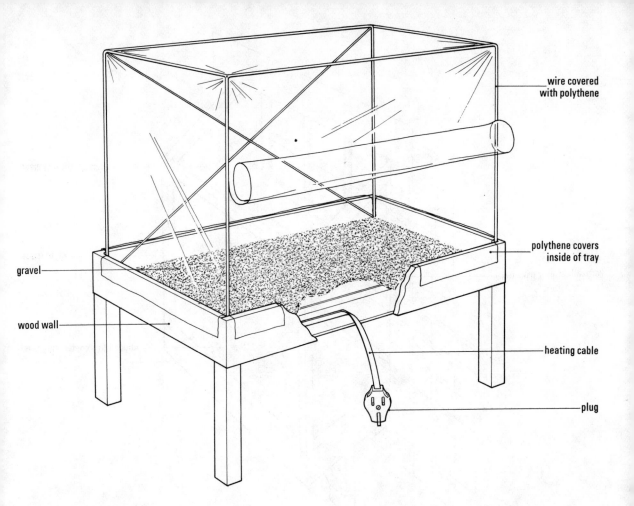

wire covered
with polythene

polythene covers
inside of tray

gravel

heating cable

wood wall

plug

Figure 43
A laboratory plant enclosure.

With some plants, regulating the period of illumination can speed up the life-cycle. For this a time-switch is useful. The kind used for aquarium lighting is suitable.

An alternative form of laboratory plant enclosure can be made by attaching a large box-like structure outside a window. The floor should be solidly built and supported, and the walls and roof should have panes of glass or of tough transparent plastic; otherwise it is constructed and used as the plant enclosure already described. The window opens either vertically or into the laboratory (*figure 44*).

A coat of whitewash over the inside surface of the glass or plastic will cut down the heating effect of the sun if it becomes excessive, without harming the growth of the plants.

While this is a more costly project, it has all the advantages of a plant enclosure kept inside the laboratory, and, at the same time, uses up no laboratory space, is less

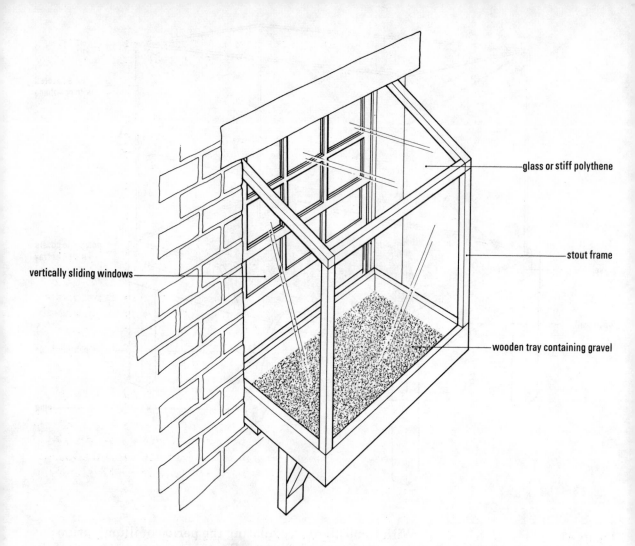

glass or stiff polythene

stout frame

wooden tray containing gravel

vertically sliding windows

Figure 44
An outside laboratory plant enclosure.

likely to be disturbed by students, and can obtain sunlight easily.

Keeping plants
Inside the enclosure plants can be kept in pots or wooden trays. A good potting and seed compost, obtainable from nurserymen and stores dealing in garden supplies, is ideal. Usually, however, a good loam soil obtained from a field or garden is suitable.

The plants should be watered as required. By dampening the gravel a reasonably humid atmosphere can be maintained.

Take care that they do not become root-bound in their containers. If this is apparent a plant should be put in a larger pot. Sometimes the plant can be broken up and propagated in several pots.

Either by adding a suitable fertilizer or by changing the soil, make sure that the plants get an adequate supply of inorganic salts.

Germinating seeds

Most seeds over 2 mm in length (or diameter) can be evenly scattered on, or buried in, soil in a wooden tray for germination. Alternatively the seeds can be uniformly distributed on masonry sand, vermiculite, or a layer of filter paper, paper handkerchief, or the like. These materials will retain moisture for some time.

Very small seeds are best germinated on damp paper or agar jelly (agar powder plus distilled water boiled to a creamy uniform consistency, poured into a container, and allowed to cool and set). An even distribution can be obtained by mixing a small quantity of sharp sand with the seeds before scattering. Petri dishes – plastic or glass – can be used as containers, preferably with the lids on to retain the moisture supply.

It is best to retain water in a container of germinating seeds with a lid of glass or transparent plastic. If watering becomes necessary a *very fine spray should be used* and the water applied gently, otherwise the seeds will be washed away or damaged. *Plants and seeds sometimes have specific requirements* for light, water, heat, and soil. This should always be checked.

In order to fit plant breeding into the school timetable, make a check of the *time taken for germination* and the length of the *period between germination and flowering* for each strain.

Figure 45
Cross-pollination by hand.

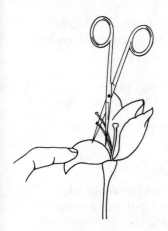
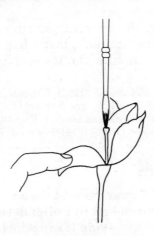

1 Stamens being cut with scissors   2 Paint brush dusting stigma   3 Polythene (transparent) bag tied over flower

Cross-pollination by hand

The agencies of pollination – insects, wind, etc. – are largely excluded from an indoor plant-growing area. With experimental breeding pollination has to be controlled whether the plants are indoors or out. In these cases pollination has to be carried out by hand.
If it is a question of merely ensuring that pollination takes place, this can easily be done by lightly flicking a camel hair brush (a small paint brush will do) within each open flower of the strain or strains that one wishes to cross. Pollen will be caught up in the brush and transferred to the stigmas of other plants.

If, however, the pollination has to be controlled, use the following procedure. A, illustrated in *figure 45*, is the flower to be pollinated, B the flower from which the pollen is to be obtained.

1 Open up A carefully while it is a bud. Remove the stamens with small forceps or scissors. Enclose A in a small paper or plastic bag tied gently round its stem.
2 *Either a* pick up pollen with a small brush from B when its stamens are ripe. Dust it onto the stigma of A, when it too is ripe, and put A back in its bag immediately after the operation.
(The brush must be thoroughly washed in alcohol, rinsed in water, and dried before it can be used for crossing other, different strains. Keeping a separate, labelled brush for each type of cross is a good idea.)
Or b place a small paper or plastic bag over B when its stamens are ripe and shake so that pollen is dispersed within it. Transfer the bag to A as a substitute for the one already enclosing it. Shake A to ensure that pollen will fall on its stigma(s). The pollen-containing bag can be left on A.
3 Attach a small tie-on label indicating the date and details of the cross.
4 When the fruit starts to develop the bag can be removed. If a transparent plastic bag is used development can often be followed without removing the bag.

Goold-Adams, D. (1969) *The cool greenhouse today.* Faber.
Hartmann, H. T., and Kester, D. E. (1968) *Plant propagation: principles and practices.* 2nd edition. Prentice-Hall.
Robinson, G. W. (1959) *The cool greenhouse.* Penguin.

Locusts

Locusts can be kept for a week or two by the method described below. Fuller details of locust culture appear in *Teachers' guide* 1, *Introducing living things*, in Chapter 9 and Appendix 3.

The perpetuation of life

Figures 46 and 47 show suitable cages. For temporary culture, the Kilner jars shown in *figure 47d*, or large jam jars, may be useful. The top of the jar should be covered with fine muslin or zinc gauze. The metal ring of the lid of a Kilner jar allows either the muslin or gauze to be held securely. About ten animals to each jar is a suitable number.

Figure 46
A locust cage.
Photograph, Peter Fry.

A small handful of fresh grass should be placed in the jar and replaced after a day or two if necessary. Care must be taken that the grass is not contaminated, particularly with insecticide. Privet and elderberry leaves can be used as alternatives. The animals should be kept at a temperature of $31 \pm 3 \,°C$. The temperature in the laboratory may be right but otherwise a lamp used as a heat source will maintain an even temperature, particularly if a shield of hardboard

Figure 47
Cylindrical locust cages.
*Photographs from Hunter-Jones, P.
(1966) 'Rearing and breeding
locusts in the laboratory', Centre for
Overseas Pest Research.*

or similar material is put behind the bulb to reflect the heat
onto the animals. The locusts are not shielded from the light.

Fuller details about cages and the care of locusts will be
found in Appendix 3 of *Teachers' guide* 1.

References	

Barrass, R. (1964) *The locust: a guide for laboratory practical work.*
Butterworth.
Hunter-Jones, P. (1966) 'Rearing and breeding locusts in the laboratory'.
Centre for Overseas Pest Research. (Available free from the Centre at
College House, Wrights Lane, London W8 5SJ.
Nuffield Biology, Revised edition (1974) *Teachers' guide* 1,
Introducing living things. Longman.

Mice

Cages

It is best to keep mice in shallow cages with separate areas for nesting and feeding. The cage described here was designed by Dr M. E. Wallace, University of Cambridge Department of Genetics. It has been found to be very suitable for use in schools.

The Cambridge cage consists of a plastic (polypropylene) bowl of the following internal dimensions: floor 28 cm × 22 cm, height 8 cm. It is covered by a wire lid sloping down towards the centre of the bowl. On the lid a metal plate acts as a cover to the nesting area. A 300 cm^3 bottle supplies water.

Figure 48
The components of a Cambridge mouse cage.
a plastic bowl
b metal division between food
c water bottle with rubber top containing a capillary tube
d metal plate of food compartment
e wire mesh lid.
Photograph, Dr C. Curds.

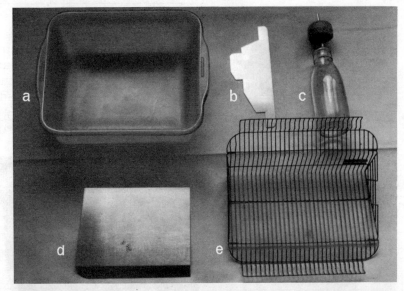

Figure 49
The mouse cage set up.
Photograph, Dr C. Curds.

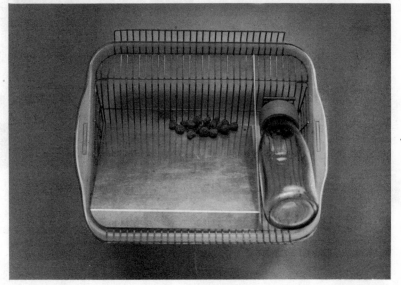

Six to nine adults can be kept in one cage. For breeding a trio of two does and a buck can be kept in the cage. Two does can nurse litters at the same time, but if the litters are of unequal age, or if one of the mothers tends to eat her young, it is better to separate the does for parturition and nursing.

Figure 50
A record card for a mouse cage.

doe *MARY 2 (se)*			parents doe *MARY 1 (se)*	buck *GEORGE 1 (se)*
date of mating	stud buck	date of birth of litter	number in litter	description of litter
14·10·73	ALAN 3 (+)	4·11·73	6	3♂+ 3♀+

Keeping cages and records
A battery of 12 cages will probably be sufficient for the course. They can be kept on racking made of metal strips fixed to a wall. A space of 16 cm is needed between shelves and a space of 5 cm between adjacent cages. Three shelves of four cages each will occupy a wall area of about 1.3 m wide by 0.7 m high and will project about 0.3 m.

Figure 51
A cage open for inspection. The nest area is on the left of the bowl. On the right, immediately below the food and drinking area, is the soiled sawdust. This need only be removed when the cage is being cleaned out.
Photograph, Dr M. E. Wallace.

A room temperature of around 20 °C and reasonable ventilation with no draughts are desirable. Each doe should have her own record card which can be clipped to the cage by the spring clip provided. If these are to be permanent records, writing in ball-point pen is a safeguard against accidental wetting. Each doe and buck used for

The perpetuation of life

mating should be given a code symbol to make it easier to construct a pedigree. The phenotypic description of litters can be kept in columns, with space for noting what is done with them and for the code numbers of those used for new matings.

Cleaning

Mice urinate and defecate during exercise and especially when eating and drinking. They are extremely clean around the nesting area. In the Cambridge cage, the soiled bedding can be scooped out without disturbing the nest. If the mice are very active, a mesh shield, shaped to fit the nesting corner, can be held over them during cleaning. With particularly lively animals, the cage can be put in a large cardboard box, a dry sink, or the equivalent. It is particularly desirable not to disarrange the nest when birth is imminent or during nursing.

If the cages are cleaned in this way, and additional bedding added once a fortnight, the smell of the mice should not be too objectionable and the animals will be healthy.

A layer of sawdust on the cage floor to a depth of about 0.5 cm (30 g) with one handful of dry hay (15 g) provides a suitable bedding. The hay should be twisted into a rough sort of nest and put in the nesting area.

It is a wise procedure to wash the bowls and other parts of the cage with hot water (75 °C) roughly once a month. If detergent or a mild disinfectant is used, the cage should be thoroughly rinsed afterwards. The plastic bowl, like the other parts of the cage, will withstand boiling water and can be autoclaved up to $68\,389\,\text{N m}^{-2}$ ($10\,\text{lbf in}^{-2}$) pressure (120 °C). This is only necessary if some serious infection should affect the battery.

Mice tend to smell in warm, unventilated conditions and in these circumstances will require more frequent cleaning.

Feeding

Rat cake pellets obtainable from many pet stores make a good food. Placed on the shelter, they are not contaminated and 1 kg can be put there. This will last at least a fortnight with a maximum cage population. Oxoid 41B diet is also suitable.

The mice obtain water from the capillary tubing inserted in the cork or cap of the water bottle and projecting about 1 cm out of it. A 2.5 cm length of capillary tubing of 6–8 mm outer diameter and 1.25–2 mm inner diameter is used. The bottle should be placed with the spout protruding through the hole below the two short horizontal bars, in the space left for it by the upright divider. It will then be at the correct angle and distance from the cage floor.

A bottle should be filled with water to 2 cm below the inner edge of the cap, the cap or cork pressed on tightly, and the bottle gently inserted into the cage. The water should be replenished when the inner projection of the capillary tube is out of water. A supply usually lasts a week.

In the Cambridge cage, mice rarely build up their bedding against the capillary tube so that flooding occurs; they may do so if there is too much bedding. A routine daily check should be made that flooding has not occurred and that there is sufficient water in the bottles.

Handling

Mice should be picked up by the base of the tail. They will stay still if allowed to hold a coat sleeve or the bars of the cage lid with their front feet; then it will be easy to examine them. They quickly get used to being handled provided this is gently done. A doe which has been handled weekly or more often from birth onwards is not likely to abandon her litter when that in turn is handled. Until the litter is 12 days old, the whole litter should be held in a warm hand, and each mouse picked up should be returned to the rest of the litter as soon as possible after examination. A doe may abandon cold young. (Sometimes, does will abandon a litter born in relatively new surroundings. However, subsequent litters born there are usually well cared for.)

When mice are first received from suppliers they sometimes find the new circumstances disturbing. At first they should only be handled occasionally, but gradually this can become more frequent. After about two or three weeks they seem to settle down.

Breeding

Mice usually become sexually mature when they are six weeks old. Gestation takes 19–21 days, longer if the doe is suckling. The minimum generation interval is about ten weeks. A generation each school term is a good general expectation.

Sexing is easiest from 0–15 days and after six weeks. The penis of the male is further from the anus than is the vulva of the female. To begin with, it is best to examine the whole litter at a few days old for this genito-anal interval. At about one week, the does show milk teats.

Social effects
Mice are social animals. They react to each other in many subtle, as well as obvious, ways. This is especially true of reproductive behaviour.

The oestrous cycle of the female is modified differently by the presence of males and by the presence of females. The cycle is shorter in the presence of a male (or of his smell) so that the female comes into heat (oestrus) once every four or five days. She will accept the male only when she is on heat.

In the presence of other females, mutual suppression of oestrus occurs for longer or shorter periods depending on the number of females living together. If such females are paired with males, the smell of the male starts a new cycle among those in which oestrus was suppressed, so that most of them come into heat on the third night.
The same reaction occurs if the male is confined behind a wire mesh grid in the box containing the females for two nights and released for the third night only, when most of the matings (about 40 per cent of females) are to be expected. The influence of the male on the oestrus cycle is known as the *Whitten effect*. It can be used, as outlined above, to regulate breeding.

Another of these subtle social reactions, known as the *Bruce effect*, is shown by the female during early pregnancy (days 1–4 after mating). The presence (or smell) of a male after the female has been removed from the stud male with which she mated may cause pregnancy to fail and the female to resume oestrous cycles, especially if the second male belongs to a different strain from that of the stud male. If the second male has access to the female, she may then mate again, this time with him. The use of males carrying different genetic markers shows that all litters born to females which have mated twice have been sired by the second male. This interesting piece of behaviour can be made use of if an unwanted or inconvenient mating occurs.

It should be realized that social effects such as the Whitten and Bruce effects are very delicate reactions and are only found among mice in good condition. They can be upset by various circumstances, some of which we still do not know.

Detecting mating

After copulation, a plug is formed in the vagina by the male. It is of a hard white substance and often completely fills the vagina so that it can be seen with the naked eye. When it is more deeply placed, it can be detected with the aid of a small spatula, gently inserted into the vagina. The plug is usually dispersed within 24 hours, but occasionally it persists for two or three days. For exact timing, examination for a vaginal plug must be made every day after the doe and buck are put together. A simple practical guide that can be used to check on the expected date of arrival of a litter is the fact that when a doe is seen from her outline to be pregnant, she usually has her litter in about a week.

In cases of doubtful paternity, the fact that the gestation period is not less than 19 days can sometimes be used to identify the sire.

Pseudo-pregnancy

Mating occasionally leads to pseudo-pregnancy, in which the oestrous cycle is in abeyance for two weeks. In this case, if possible, make the cross again, with another pair of animals.

Preserving offspring

If facilities do not allow offspring to be kept alive, these can be killed and preserved by deep freezing. When the results are being analysed, the preserved mice can be brought out for scoring.

A collection of preserved mice can also be used for setting problems. Thus, the offspring of several crosses can be set out and students asked to work out the likely genotypes of the parents.

Wallace, M. E. (1971) *Learning genetics with mice.* Heinemann Educational.

Planaria, culture of

Planarians are best kept in pond water in a shallow dish. Tap water has been proved to be suitable by some authorities. The conditions should be cold and dark. The water should be changed two or three times a week and the dish should be rinsed out.

While regeneration is taking place the animals are not fed but, at other stages, they should be given a small piece of liver two or three times a week. After half an hour, remove the uneaten food. This can be done at the same time as the water is changed. At all times make sure that the water does not become contaminated.

The following British species are suitable for regeneration work.

Dendrocoelum lacteum: very common in streams and standing water.
Planaria vitta: found in the mud of wells and springs.
Planaria lugubris: found in gently running streams and still water.
Polycelis nigra: found in slow streams and standing water.

These species can be found throughout the year. *Dendrocoelum lacteum*, *Planaria vitta*, and *Planaria lugubris* have marked natural powers of regeneration and can be observed to tear themselves in two on occasions. It is well to point this out to students so that they realize the experimental work on regeneration is based on a natural phenomenon, and hence it is not inflicting harm on the animals.

Of the four species mentioned, *Polycelis nigra* is possibly the easiest to keep in the laboratory.

Planaria can sometimes be obtained from biological supply agencies.

Roots

The plants listed below will provide suitable material for root tip squash preparations. Onion is strongly recommended.

The times recommended for collecting, in the righthand column, relate to plants grown out of doors in the south of England. (In northern areas, development may be delayed in the early months of the year.)

Plants		Approximate time recommended for collecting
Allium cepa (onion)	$2n = 16$	February–June
Allium sativum (garlic)	$2n = 16$	February–June
Allium schoenoprasum (chives)	$2n = 16$	April–September
Crepis spp.	$2n = 16$	March–June
Endymion non-scriptus	$2n = 16$	January–March
Fritillaria spp.	$2n = 24$	March–November
Hyacinthus spp.	$2n = 16$	January–April
Lilium spp.	$2n = 16$	June
Trillium spp.	$2n = 10$	April–September
Tulbaghia spp.	$2n = 12$	Foreign; keep in greenhouse or cold frame
Vicia faba	$2n = 12$	March–April

The roots should be in fresh, growing condition.

Bulbs and seeds can often be induced to produce roots in the laboratory all the year round. Arrange bulbs with their bases sitting on water, by supporting them in a beaker with matches or toothpicks. Special vases for this purpose are available but not really necessary. Keep the bulbs in the dark or in subdued light.

Figure 52
Supporting a bulb in water to encourage root development.

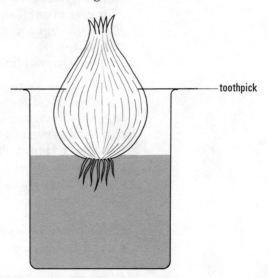

toothpick

Bulbs tend to become dormant during the autumn and winter. This can be overcome by keeping them in a refrigerator for 4–6 weeks. If they are then treated as shown above, roots should be produced.

Seeds of *Vicia faba* can be germinated all the year round and will produce suitable root material. The following procedure can also be adopted for use:

Soak the seeds in water for 24–36 hours. Remove the cracked testas and plant the seeds in damp vermiculite. After two days when the radicle is about 2 cm long, remove 2–3 mm from its tip. This will stimulate growth of lateral roots. After a week in damp, but not flooded, vermiculite a crop of 15 beans should provide sufficient roots for a class of 30.

With plants kept in pots care should be taken that they do not get root bound, since this retards root development. If cuttings are taken and planted in fresh pots containing John Innes No. 1 compost, fresh root growth is stimulated. *Tulbaghia* species (Liliaceae) kept in a cold frame or greenhouse respond well to this treatment and can provide material during most of the year.

More satisfactory preparations can usually be made if the

The perpetuation of life

roots are fixed in acetic alcohol (1 part acetic acid : 3 parts absolute alcohol) for a day or so, after removing them from the plant. They may be stored in this fixative, in a refrigerator, for several months. In this way roots of some plants can be used at times when they would otherwise be unobtainable.

Specimens, preservation of

Drying

Some plant and animal material can be dried. This is probably best done in an oven at a temperature that will not cause burning (about 40 °C), or in a desiccator. Indicator type silica gel is a good drying agent.

Pressing

Thin plants can be pressed – laid between sheets of absorbent paper, e.g. blotting paper, with weights put on top. Books or bricks will do as weights. The material must be carefully laid out before the upper layer of paper is put on and should be tight in the press but not crushed or broken. It should be left for about a fortnight. Pressed material is suitable for work in Chapter 1.

Liquid preservatives

If material is kept in liquid preservatives (70 per cent alcohol or 5 per cent formalin are the most common), it should be taken out and washed in tap water before use in class. If examined on absorbent paper, e.g. filter paper, the operation need not be messy. Provided the material is returned to the preservative immediately after use it need not be significantly damaged.

Preserving the colour of plant material

An easy method which gives good results but which tends to exaggerate the green colour is described below.
1 Put the material in a solution consisting of:
90 parts of 70 per cent alcohol
5 parts of 40 per cent formalin
5 parts of glacial acetic acid
2 Add copper sulphate crystals. Experience will show how much is needed for a particular species. Little harm will be done by excess.
3 Store the material in the solution or transfer it after 48 hours to 70 per cent alcohol for storing.

Other, more reliable, methods are:

For overall colour preservation

20 g	phenol
20 g	lactic acid, Sp.GT 1.21
40 g	glycerine, Sp.GT 1.25
20 cm³	distilled water
0.2 g	cupric chloride
0.2 g	cupric acetate

The specimens can be dropped into this fluid and stored until needed. Alternatively they can be transferred after about ten days to 2 per cent formalin.

For preserving green colour

a Obtain a saturated solution of copper acetate in 50 per cent acetic acid.
b Dilute to 1 in 2 or 1 in 4 with distilled water before use. The lower strength is necessary for delicate specimens.
c Bring to the boil and drop in the specimens. Boil for 5–15 minutes. Tough material will need longer boiling.
d Wash for several hours in running water – tap water will do.
e Store in 3 per cent formalin.
Sometimes soft fleshy fruits will burst and some plants will shed their leaves in this preservative, which is nevertheless suitable for a wide range of material.

For preserving red colour

a Dissolve 100 g of zinc chloride in 2000 cm³ of hot distilled water. It is unnecessary to boil.
b Filter the solution while hot.
c Add 50 cm³ of glycerin and 50 cm³ of 40 per cent formalin to the filtrate. If a precipitate forms, decant on cooling.
d Place specimens in clear solution and leave.

Freezing

Many organisms can be preserved in a refrigerator. For periods of over a week or so, a freezer compartment, or better still a deep-freeze as used for ice cream, is necessary. Small mammals, such as mice, are very suitably preserved in this way. If they are carefully dried when thawed they have a lifelike appearance and are most suitable for exercises in which differences and similarities are sorted out.

Cleaning bones

Bones obtained from carcasses to be used for studying limbs, skulls, etc., should have as much flesh as possible removed from them with a knife and needle. They can be cleaned in a number of ways.

1 Leave them buried in the ground for 2–3 months. The flesh will be removed by scavengers and by decomposition.
2 Leave the bones in cold water. The flesh will come off by decomposition. This method produces a smell after a while. It takes several weeks.
3 Keep the bones in boiling water for about two hours. This also is a smelly operation. The softened flesh is easily scraped off the bone afterwards.
4 Use the following solution:
   5 dm^3 tap water
   20 g   sodium sulphide
   20 g   Pancreatin
Place the bones in the solution, warm it, and continue until the flesh has either come off or is easily scraped off.

After cleaning, wash the bones well in cold water to which a little hydrogen peroxide has been added. This will bleach the bones which will acquire a suitably clean and white appearance especially if they are dried in sunlight.

Index

A

Acetabularia, experiments on development of, 80, 81
acetic alcohol, 193, 221
adaptation, 62–3
adaptive radiation, in mammals, 171, 173
agglutinins, 10, 11
agglutinogens, 10
air pollution, and distribution of peppered moth, 153
albinism, 73; in plants, 85
alleles, 143
Allison, A.C., 156
Aloe: meiosis in formation of pollen in, 51–5; mitosis in root tips of, 46–7
amino acids: produced in Miller's pyrosynthometer, 41; triplet code in DNA for, 75
Amoeba, exchange of nuclei of, 81
amphibians: induction of breeding in, 197–8; metamorphosis in, 79, 91–3; observation of development of, 105; rearing eggs and tadpoles of, 199–200
anaemia, sickle-cell, 156–7
anatomy, comparative: evidence for evolution from, 173–4
animals, law relating to experiments on, 196–7
anterior–posterior gradient, in regeneration of planarians, 114
antibiotic discs, 65
antibiotics, mutant bacteria resistant to, 65–8
antibodies, 14; to blood group substances, 10, 11
antigens, blood group, 10
Antirrhinum majus: experiments on inheritance in, 35; mutations in, 62
apes, compared with humans, 182, 184, 186, 189
Arabidopsis thaliana, mutations in, 63–4
arginine, metabolic pathway for synthesis of, 74, 75

armadillo, nine-banded, 80
Asparagus, self- and cross-fertilization in, 119
Aspergillus, conidiophores of, 121
Australian continent, isolation of, 172
Australopithecus, 181, 184, 186, 188–9
autoclave, 66
auxins, 87–9; in weedkillers, 89

B

Bacillus cereus, 66
Bacillus subtilis: culture medium for, 194; for study of mutation, 65, 66
bacteria: antibiotic-resistant mutants of, 65–8; binary fission of, 121
barley (*Hordeum vulgaris*): breeding experiments with, 35; development of seedlings of, 86; measurements of growth of seedlings of, 107–8
beads, models using, 30–1, 142–9, 163–5
Beagle, H.M.S., voyage of, 170
Begonia, leaf cuttings of, 116
biomass, 102
bipedal locomotion, in humans, 184–5
birds: courtship in, 127; identification of, 7; selection by, 155–6, 157–9
birth control, methods of, 135–6
birth rate, England and Wales (1930–70), 132–3
blood: obtaining samples of, 12–13; transfusions of, 10, 11; *see also* sickle cells
blood groups, 10–12; testing for, 12–14
Blood Transfusion Service, National, 12
bones, methods for cleaning, 222–3
brachydactyly, 73
Bryophyllum, leaf plantlets of, 121

bulbs, obtaining roots from, 220

C

cabbage, mutation in, 64
camouflage, 159; experiment imitating, 160–3
cancer, 117
castration, of cattle, 90
cattle: artificial insemination of, 166; castration of, 90; rearing of, for beef and for milk, 83
cell division: graph of, 106; in growing tissues, *see* mitosis; in making gametes, *see* meiosis; and pattern of development, 105–9
cells: cultured *in vitro*, 113; from different organisms, fusion of, 82; differentiation of, 109–13; interaction of, 114
centromeres, 46–7
cerebral cortex, of human brain, 180
characteristics of organisms, 5; inheritance of, 24; new, produced by mutation, 71; selection of, for comparison, in plants, 6–7; variation of, within a species, 5, 9
chimeras, 82
chloramphenicol, 65
chorionic gonadotrophin, for induction of breeding in *Xenopus*, 197
chromatids, 47, 55
chromosomes, 45; aberrations in, 69–70, 71; as basis of inheritance, 56; of humans, 72–3; multiplication of (polyploidy), 70–1; and mutations, 69, 71–2; numbers of, in different organisms, 70; rate of, 69
Chrysanthemum, mutation in, 64
classification of organisms: of humans, 179, 181–2: international system of, 7–8; natural and artificial, 18

co-dominance, in inheritance, 39
colchicine, as mutagenic agent, 72
coloration, as camouflage or as warning, 159
colour-blindness, inheritance of, 137–8, 161
contraception, 135–6
contraceptive pills, 132, 135–6
couch grass, rhizomes of, 121
courtship, 125–8
cranial capacity: changes in, in evolution of humans, 182–3
cress seedlings, auxin and development of, 87–8
cucumber, inheritance of cotyledon flavour in, 38
culture media: for bacteria, 194; for *Drosophila*, 25
cuttings: of *Begonia* leaves, 116, 122; of geraniums, 121–2
cytoplasm: and nucleus, in development, 80–2

D

Darwinism, deductive method of, 170–1
DDT, inheritance of resistance to, 68
de-differentiation, 116
deductive method, of Darwinism, 170–1
development: cell division and, 105–9; control of, in whole organism, 114–16; definition of, 79–80; different patterns of, 105; inheritance and environment in, 24, 82–6; measuring variation in, 96–101; nucleus and cytoplasm in, 80–2; problems on, 101–2; rules for verifying hypothesis that a substance affects, 94–5
differentiation of cells, 109–13
dinosaurs, 171
DNA, 179; triplet code for amino acids in, 75
dogs, breeding of, 120–1
dominance, in inheritance, 30, 33, 120
Drosophila: breeding experiments with, 21–2, 24–9; culture of, 200–5; insecticide-resistance in, 69; mutations in, 63; sexing of, 205
dwarfs and giants, 96–7

E

eggs, of *Pomatoceros*, 44
Eldon blood-group cards, 12
embryology, comparative: evidence for evolution from, 179

environment: and inheritance, in development, 24, 82–6
enzymes, one-to-one relation between genes and, 74
Escherichia coli, isolation of streptomycin-resistant strains of, 68
ether–ethanol mixture, for anaesthetizing *Drosophila*, 24
etherizer, 24
ethnology, blood groups in, 11, 14
evolution, 141, 170; convergent, 173; evidence for, 171–9; of humans, 179–81; problems in, 190
evolution game, 173
experiments: controls for, 23, 41, 89, 91; on live animals, law relating to, 196–7
exponential series, 109
eye colour, inheritance of, 9, 36, 37, 39
eyes, of primates, 180

F

family tree: construction of, 36; of Wedgwoods, Darwins, and Galtons, 24
fertilization, 119; in *Pomatoceros*, 21; self- and cross-, in plants, 119–20, 124
Feulgen stain (Schiff's reagent), 44, 109, 193
finches of Galápagos Islands, 172–3
finger prints, 10, 15–16; classification of, 16–18
foot, human, 187
fossils, evidence for evolution from, 174–8
frogs: development of, 105; experiments on eggs of, 82; feeding of, 200; induction of breeding in, 197; thyroid gland and metamorphosis of, 91–3
fungi, reproduction of, 121

G

Galápagos Islands, 170, 172
Galton, Sir Francis, 16
gametes, 21, 55; cell division in making of, *see* meiosis
garden, genetic, 206
gene pool, in a population, 143, 145
genera, 8
genes: mutations in, 71; one-to-one relation between enzymes and, 74; in a population, 141–2; symbols used for, 33–4
genetic counselling, 137
genetics, evidence for evolution from, 179

genotype, 29; selection and, 152, 156, 165
geographical distribution of organisms, evidence for evolution from, 171–3
giants and dwarfs, 96–7
gradient plate technique, for isolation of bacterial mutants, 67
grafting of roses, 122–3
graphs: of cell division, 106; of growth of barley seedlings, 107
growth, definitions of, 79–80

H

haemophilia, 62, 73
hair colour, inheritance of, 9, 39
Hammerling, Prof. J., 80
hand: evolution of, and use of tools, 188–9; power and precision grips of, 188; of primates, 180–1
Hardy–Weinberg equation, 142, 149–51
Henry, Sir Edward, 16
Herschel, Sir William, 16
heterozygote, 27, 142; in absence of dominance, 33; may have advantage over both homozygotes, 156–7
Holtfreter's medium, 193, 199
Homo erectus, 181, 183, 184, 189
Homo sapiens neanderthalensis (Neanderthal man), 184, 189
Homo sapiens rhodesiensis (Rhodesian man), 184
Homo sapiens sapiens (modern man), 184; compared with apes, 182, 184, 186, 189; *see also* humans
homologies, 173, 174
homozygote, 27, 142
hormones, animal: and development, 90–4
hormones, plant: practical use of, 89
horse, evolution of, 174–8
houseflies, insecticide-resistance in, 68–9
humans: chimeras with cells from, 82; chromosomes of, 72–3; classification of, 179, 181–2; control of breeding of, 132–7; courtship in, 127–8; development of, 102, 103; evolution of, 179–89; measurements of growth of, 102, 103; special qualities of, 183, 189

I

illegitimacy, statistics of, 135
immunity, 14